Beginning Analog Electronics Through Projects

Second Edition

Beginning Analog Electronics Through Projects

Second Edition

Andrew Singmin

Newnes

Boston Oxford Johannesburg Melbourne New Delhi

Newnes is an imprint of Butterworth–Heinemann.

Copyright © 2001 by Butterworth–Heinemann

A member of the Reed Elsevier group

All rights reserved.

No part of this publication may be reproduced, stored in a retrieval system, or transmitted in any form or by any means, electronic, mechanical, photocopying, recording, or otherwise, without the prior written permission of the publisher.

Recognizing the importance of preserving what has been written, Butterworth–Heinemann prints its books on acid-free paper whenever possible.

 Butterworth-Heinemann supports the efforts of American Forests and the Global ReLeaf program in its campaign for the betterment of trees, forests, and our environment.

Library of Congress Cataloging-in-Publication Data
Singmin, Andrew, 1945–
    Beginning analog electronics through projects / Andrew Singmin.—2nd ed.
      p. cm.
    New ed. of: Beginning electronics through projects / Andrew Singmin. c1997.
    Includes index.
    ISBN 0-7506-7283-8 (pbk. : alk. paper)
    1. Electronics—Amateurs' manuals. 2. Analog electronic systems—Amateurs' manuals.
    I. Singmin, Andrew, 1945– Beginning electronics through projects. II. Title.
    TK9965 .S544 2000
    621.381—dc21
                                                                               00-063805

British Library Cataloguing-in-Publication Data
A catalogue record for this book is available from the British Library.

The publisher offers special discounts on bulk orders of this book.
For information, please contact:

Manager of Special Sales
Butterworth–Heinemann
225 Wildwood Avenue
Woburn, MA 01801-2041
Tel: 781-904-2500
Fax: 781-904-2620

For information on all Newnes publications available, contact our World Wide Web home page at: http://www.newnespress.com

10 9 8 7 6 5 4 3 2 1

Transferred to digital printing 2006

For a special someone

Contents

Preface ix

Introduction xi

1 **Overview of Analog and Digital Electronics** 1

2 **Resistors** 7

3 **Potentiometers** 11

4 **Multimeters** 13

5 **Ohm's Law** 15

6 **Light-Emitting Diodes** 17

7 **Switches** 19

8 **Capacitors** 21

9 **Integrated Circuits** 23

10 **Tools** 25

11 **Soldering** 27

12 **Basic Electronics Theory** 29

13 **Reading Schematics** 31

14 Assembly Techniques   33

15 Handling Components   37

16 The SINGMIN PCB Circuit Assembly Board   39

17 Construction Details for 10 Simple Projects   43

Project #1: Fixed Low-Frequency LED Flasher   43
Project #2: Variable Low-Frequency LED Flasher/Driver   52
Project #3: Fixed Low-Gain Audio Power Amplifier   59
Project #4: Fixed-Frequency Audio Tone Generator   65
Project #5: Variable-Gain Audio Power Amplifier   71
Project #6: Fixed-Gain Audio Preamplifier   77
Project #7: Guitar Headphone Amplifier   83
Project #8: Visual Electronic Metronome   89
Project #9: Variable-Gain, Hi/Lo Response Audio Preamplifier   94
Project #10: Dual-Gain Electret Microphone Audio Preamplifier   101

18 Troubleshooting Test Equipment   109

Project #11: Signal Injector   109
Project #12: Signal Monitor   114

Index   119

Preface

Much has transpired between the release of my first edition of *Beginning Electronics Through Projects*, published in 1997, and this second edition, *Beginning Analog Electronics Through Projects*. For the most part, I've been busy developing several new books, including *Practical Audio Amplifier Circuit Projects* and *Beginning Digital Electronics Through Projects*. You'll find many significant differences between the first edition of this book and the current edition, many of which you'll need to know to proceed with the projects. Please refer to the introduction for a list of the key changes.

There will no doubt always be a succession of new-generation enthusiasts embarking on their first introduction to electronics (be it for fun or work), and on the lookout for easy-to-understand introductory books on the subject. For those beginner enthusiasts and hobbyists, my books are suited to your needs.

Have fun building the projects and, at the same time, increasing your working knowledge of electronics.

I had fun working with you, Candy!

Andrew Singmin, 2000

Introduction

This book is divided into two main sections: The first part (Chapters 1 through 16) covers some of the basic descriptions and theory of electronic components, and the second part (Chapters 17 and 18) provides detailed instructions for constructing 12 simple electronic projects. The initial section deals exclusively with only the types of components that are later used in the projects.

The beginner in electronics is often handicapped by exposure to far too many subject areas, even at the elementary level. Thus textbooks traditionally have the habit, in the interest of completeness, of covering so many subject topics that, more often than not, the beginner is left confused. I speak from experience, remembering my own start in electronics!

This book, therefore, takes a different approach in that only the components and theory that are of use and relevance to the chapter on project construction will be covered. The aim here is to make the beginner comfortable with a limited range of components and circuit theory, rather than confused over a wider range of topics.

The emphasis in this book is focused on the practical aspects of electronics. Through the construction of the projects and a developed understanding of the theory behind the components and techniques, I hope you will gain a better appreciation of the subject and perhaps be encouraged to pursue electronics even further. When I started in electronics as a hobby, I experienced the tremendous shortage of good, practical books on the subject. Those few that were available were either too theoretical (most of them), described circuits that had components that were impossible to obtain, or were far too difficult to understand, let alone build. Many beginners take on projects that are far too complex and become discouraged when the circuit fails to work; however, by focusing here on simpler circuits to begin with, the chances of success are much higher, especially in view of the abundance of clear instructions given.

All of the electronic components and basic electronic techniques described in these early chapters will be put to good use later in building the 12 construction projects. It will be extremely beneficial for you to work through these early chapters carefully, making sure you understand things as you progress. No matter how complex electronic circuits are, they can always be broken down into simpler parts that are easier to understand. Although many more different components exist than the types listed here, from a beginner's point of view we are limiting the description to just those components that you will encounter in the construction projects. We will go through just as much theory as is needed to cover the construction projects. In short, everything that follows is relevant.

Armed with this book, a collection of electronic components, and a hot soldering iron fired up and ready to go, you should be well on your way to the start of an exciting new hobby or even on the threshold of a new career!

Please note the following changes between the first edition of this book and the current edition before proceeding:

1. The projects in the first edition were designed to use a dedicated prototype board of my own design called the SINGMIN PCB. This board was developed in order to overcome the limitations found with existing assembly boards. When the first edition of this book was written, I used the SINGMIN PCB extensively for circuit builds, so the projects in that book showcased this unique design.

    Several years have passed since then and my custom assembly platform is no longer available. It is not mandatory, however, to use this specific PCB, and you can use whatever alternate assembly boards suit your own preferences. Alternately, you can use the design of the SINGMIN PCB as an example template or layout guide to design your own boards, using whatever commercial assembly is available. The two extra projects in this second edition (#11 and #12) are left for you as an exercise in initiative—to do the board layout yourself. Here are some useful placement tips when either using a universal prototyping board or putting your own board design together.

    - Assign a power rail running horizontally across the top of your board to be the positive voltage rail.
    - Assign a ground rail running horizontally across the bottom of the board to be the negative (or ground) rail.
    - Because power connections will likely be required at the bottom of the board, also assign a secondary power rail along the bottom of the board. This will be connected to the upper principal power rail.
    - A similar premise applies to the ground rail, so add a secondary ground rail along the top of the board, and have it connected to the principal lower ground rail.
    - Allow plenty of space for the Vcc/2 bias components. You will always need these when an operational amplifier is

operated off a single (9-volt) supply. You will notice in all of my circuits of this type that several of these components are needed. These components (nominally two resistors and a capacitor) can consume a considerable amount of not only board space, but also valuable solder connections.
- Keep the component layout as close as possible to the schematic layout, so troubleshooting is simplified when you're verifying wiring interconnection traces and component placements.

2. The project kits that were mentioned in the first edition of this book are no longer available. The components, however, are all readily available at your local electronics component store.

Now it's time to begin with Chapter 1. If you like to work to music like me, put your favorite tape or CD on (for me, it's the Dixie Chicks), turn up the volume, and let's start building.

CHAPTER **1**

Overview of Analog and Digital Electronics

Analog Electronics

Sooner or later you'll ask, what's the difference between analog and digital electronics? Analog circuits are conventionally where most beginners to electronics start because we can identify much more easily with analog concepts in electronics. Take, for example, the simplest and best analog scenario—audio amplifiers. Everyone is familiar with audio amplifiers in one form or another, from TV, stereo sound systems, cassette recorders, Walkmans, CD players, and so on. Audio is a concept we can readily identify with. You can be guaranteed that almost all people, at some point in their lives, have encountered audio. If our ears pick up sound, we know it's there (so to speak).

We can distinguish easily between clear, clean sounds and distorted sounds, low sound levels and high sound levels. Turn up the bass on your stereo system and everyone knows (you don't have to be electronics-knowledgeable) what that sounds like. Jimi Hendrix's screaming guitar work (perhaps not to everyone's taste) is clearly different from a classical Mozart rendition. Once more, no electronics appreciation is needed to perceive the difference.

That's why analog electronics is so satisfying to work with—we can see (well, actually hear) the results and appreciate the differences instantly. Video circuits, although still analog, are much more difficult to work with. There are no equivalent easy-to-use circuits as we have in the audio case, and even if there were, they would still not be the same clarity in being able to distinguish between different video characteristics. Video gets a little far removed from the user being able to distinguish differences. Why do I mention video here? Well, digital is still one more step removed than video and is, therefore (in my opinion), an even more faceless entity—and therefore a more difficult concept for the beginner to appreciate.

Temperature is another entity that is an analog parameter. We all are familiar with reading temperatures, from the thermometer on your wall used to monitor room temperature changes, to taking your temperature with a medical thermometer. The temperature reading changes smoothly; there is no jump from one level to another. Depending on the type of thermometer deployed, the incremental changes in temperature will be different, but it is always continuous. That's one of the key properties of an analog signal—it's continuously varying.

We (i.e., humans) are familiar with entities that change in this manner. Analog electronics is concerned with the processing of analog signals, which is usually amplification, filtering, or signal generation. And once more, the electronic amplifier is the prime example of analog electronics. Most beginners' introduction to electronics (and it is usually analog electronics) is the encounter of the basic amplifier block. Amplifiers, by implication of the title, change a small signal into a larger signal.

Let's say we monitor (by one means or another) the starting signal, pass it through an amplifier, and then look at the resultant signal; if it has increased in magnitude, then we know it's been amplified. Anything from preamplifier circuits, which generally increase the voltage level (but not the current level) of the starting signal, to power amplifiers, which increase the drive (i.e., current) capacity of the signal, is included in this category. Preamplifiers, although increasing the signal amplitude, don't have any capability to drive a low-impedance load, such as a speaker, and many applications (e.g., audio systems) have a final requirement of needing to do that.

Where we're going to ultimately drive a speaker (this is a low-impedance load, meaning that the resistance of the load is low, typically less than 10 ohms), we're going to consume significant current. From Ohm's law we know that when a voltage is applied to a resistance, the resultant current flow increases in inverse proportion to the load resistance. A low load resistance draws more current, which in turn means that the power amplifier driving this load must be capable of supplying the load current. That's the principal difference between a preamplifier (which is essentially a voltage amplifier) and a power amplifier (although it has some voltage gain, too), which is essentially a current-supplying (or power) amplifier.

Analog amplifier circuits are always characterized by having the input signal and output signal capacitively coupled. The presence of the capacitors removes the dc component of the signal, and passes only the ac component for processing (i.e., amplification in this case), since the information we're interested in is contained within the ac signal. Analog signals are typically sinusoidal in form, which means that apart from the dimension of amplitude, they are also characterized by a dimension of frequency. Frequency is the measure of the number of complete cycles of the sine wave that occurs in one second, where the second is defined as the period of reference. A low-frequency audio signal, for example, could be 100 Hz, which means that over a 1-second duration there are 100 complete cycles.

For a sine wave, a cycle is defined as the difference between a reference point on the waveform to the same reference point on the next cycle. A sine wave basically makes an excursion from zero volts, up to a maximum, back to zero volts, down to a minimum, and back to zero again; this excursion is defined as one cycle. The waveform just repeats in time and is defined as a periodic waveform. This waveform can be easily seen by coupling it into an oscilloscope, which is basically an instrument for monitoring periodic waveforms. A high-frequency signal (e.g., 10 kHz) would have more cycles contained in a 1-second reference measure—10,000 cycles to be exact. The reference measure (whatever it is) of course needs to be defined in order for a measure of the signal of interest to be made.

The most popular and commonly used analog circuit today is undoubtedly the operational amplifier, which is most generally encountered in the inverting amplifying mode. Amplifier circuits are one of the most common circuits seen in all of electronics because all electronic signals of interest, such as microphone signals or radio signals, are very weak (i.e., small in amplitude), and hence need to be amplified to a useful level. Using the operational amplifier as the basis of construction, it is extremely easy to design and build a stable circuit.

That's the beauty of operational amplifier circuits—the circuit performance is predictable and unaffected, in the main by the operational amplifier itself. The operational amplifier gain is just determined by the ratio of two resistors. Contrast this with building up an amplifier from discrete circuits (transistors, resistors, and capacitors), where the circuit performance is going to be affected by the transistor characteristics and a mix of the circuit components used.

As your electronics expertise and enthusiasm grow through the practical learning process of learning by doing, you can move on to other circuit projects at a more intermediate, but still understandable, level. You'll find these projects in my publication titled *Practical Audio Amplifier Circuit Projects*, where there are 16 audio-based projects included, in addition to a wealth of more detailed information provided on audio amplifiers and audio-related circuits.

Because I find a lot of enjoyment running through blues riffs on a Fender Stratocaster, my emphasis is on electric guitar music projects. So that book's highlight is a nice Guitar Pacer project, which allows you to jam alongside your favorite musician because this unit mixes in your own guitar work with the tape track of your guitar idol.

At this stage, too, you might also be looking for a dictionary of electronics terminology. Check out another one of my books, *Dictionary of Modern Electronics Technology*, where you'll find a good cross-section of modern terminology used in electronics. I've included what I've found in all my years of electronics activity, both at work and for fun, and all the terms that I needed to find the definitions of.

Digital Electronics

Digital electronics, and especially digital signals, on the other hand, are quite different (from analog signals). The amplitude excursion (for digital signals) is no longer continuous, but discrete; that is, moving between two clearly defined levels, which are generally defined as the starting zero voltage point and a positive maximum. The digital signal is also periodic (like the analog sine wave), but instead of being an essentially smoothly moving continuous signal, it just goes from zero level to a positive maximum level and back to zero again. This is like a series of sharp quantum changes in signal amplitude. This sharp change can also be discerned in an audio sense.

If you were to listen to a digital square wave (which is what it's commonly known as), then the sound would be rough and harsh. The harshness can be thought of as being aligned with the quantum nature of the digital signal. On the other hand, the analog counterpart sounds smooth—an artifact that can be associated with the inherent continuum of the analog waveform's property.

The same time reference measure (one second) is used in order to characterize the digital signal of interest. Digital signals are defined by repetition rate rather than by frequency, but it is essentially the same thing. A 1-kHz repetition rate digital signal has 1,000 cycles (defined in exactly the same way as the analog signal) per second, whereas a 10-kHz digital signal has 10,000 cycles contained in a 1-second measure. The oscilloscope can also provide a visual display of a digital signal because it is also a periodic signal (regardless of waveform shape).

Digital electronics are not involved with amplification or filtering of signals (as in the analog case), but rather concern issues such as counting pulses and triggering subsequent circuits when a number of predetermined pulses have been generated. A pulse is just a single digital signal going from a zero value, to a maximum, and returning to a zero value, as opposed to the alternate, which is a train of continuous pulses. The occurrence of a pulse event is based on detecting either the positive-going pulse edge (where the pulse goes from a zero voltage level to a positive maximum voltage level), which is the more usual case, or alternately, the negative-going edge (where the pulse goes from a maximum voltage to a zero voltage level) of the input digital signal.

Digital circuits are also known as logic circuits because they essentially traverse two logic states, defined as either a logic low state (a descriptor of when the digital signal sits at the zero volts level) or a logic high state (a descriptor of when the signal sits at the positive maximum volts level). In a typical counting application, we would see a digital circuit performing a sequential count on the number of pulses being generated. When the count total reaches a predetermined value, a single resultant pulse is generated, and this pulse can then act as a trigger for a further series of logic events to take place. Digital circuits are typically a sequence of logic events taking place. Because a digital signal has a clean edge (these are the rising and falling edges), it is easy for logic circuits to respond to the edge change and, consequently, to produce a resultant pulse. Analog signals don't have this clean characteristic edge.

There are two main types of digital circuits popularly in use: TTL digital and CMOS digital circuits. The first of these circuits, TTL (transistor transistor logic) technology, is characterized by having a power supply voltage that operates off 5 volts. This technology is the most commonly used digital type and can be found in most logic designs. Although the digital schematic is essentially transparent to the technology type used, the parts list will define the technology type. The second digital technology type in use is CMOS (complementary metal oxide semiconductor) technology, which runs off a supply voltage range from 3 volts to 15 volts. But the main distinction between CMOS and TTL is the considerably lower power supply current requirements of CMOS. The consequent lower power consumption (for CMOS) translates to a longer, more desirable operating life (from battery-powered circuits). A continuous drive toward a lower supply voltage means a more compact battery requirement.

I've found that digital electronics books for beginners tend to be less successful than their analog counterparts in making the presentation both interesting and informative, because of the more abstract nature of digital electronics. Trying to get excited about binary digits and truth tables is not much to work with. So the approach I've taken, in my excursion to cover this topic, is to follow the same successful format I've used for my analog books, which is to focus on a learning-by-building approach. With the right approach, almost any topic learning task can be made fun.

Using project builds to learn has always been fun for me, so I'm guessing it's going to be fun for you, too. My *Beginning Digital Electronics Through Projects* book has a central focus of learning by building digital-based circuits that are simple to follow, use readily available components, and limit the complexity to just a single integrated circuit. This is not to say that the circuits described have limited value—far from it. You don't need a complex array of ICs for a circuit to be innovative and useful. This is a useful book to get a solid grounding in the world of digital electronics. For fun, there's an FM transmitter circuit project that's easy to build and fun to use. It's RF and not digital, but it's a fun circuit and, as the author, I kind of like having the freedom to include it—so that's why it's there!

CHAPTER **2**

Resistors

Resistors are not only the most common type of component found in electronics construction projects, but they are also the simplest to use. A resistor can be recognized as a small, tubular-shaped component with a wire lead coming from each end and a series of color bands on the body. Electrical resistance is measured in units of ohm, with higher values being preceded by the word *kilo* for thousands of ohm, thus 10 kohm is equivalent to 10,000 ohm.

A resistor's resistance value is deciphered with the color code shown as follows. Each color corresponds to a particular number. It is advisable to take time to learn the color code.

Color Band	Equivalent Number Code
Black	0
Brown	1
Red	2
Orange	3
Yellow	4
Green	5
Blue	6
Violet	7
Gray	8
White	9

Basically, three color bands are used to represent the resistor value. For example, a 100-ohm resistor has the color bands brown, black, brown. The first band color (brown) has a value of 1', the second band color (black) has a value of 0', and the last color (brown) has a value of 1'. The first two colors always represent the first and second numbers of the resistor value, and the third color indicates the number of zeros needed after the second number. Thus we have

brown, black, brown = 100 ohm

7

By going through a few more examples, you will soon see the pattern emerging. A 27-kohm resistor is represented by the color code

$$\text{red, violet, orange} = 27{,}000\,\text{ohm} = 27\,\text{kohm}$$

A 47-kohm resistor has the color bands

$$\text{yellow, violet, orange} = 47{,}000\,\text{ohm} = 47\,\text{kohm}$$

One thousand ohm is represented by the letter k; that is, 1,000 ohm is the same as 1 kohm, and 27,000 ohm is the same as 27 kohm. A value such as 2,700 ohm is therefore the same as 2.7 kohm. For even larger values, the letter M represents one million ohm. Thus, 1,000,000 equals one million ohm or 1 Mohm.

In dealing with resistors for construction projects, it is necessary to know the color code in order to select the correct resistor value. You should be familiar with translating color codes to resistor values and, conversely, resistor values to color codes.

At this stage, we need to introduce a fourth color band. This band, typically either gold or silver, represents the tolerance of the resistor, gold being a 5-percent tolerance resistor and silver a 10-percent tolerance resistor. Tolerance relates to the allowable spread in resistance value. Assume that we have a 100-ohm resistor, so the color bands are brown, black, brown, with a fourth gold (5 percent) band. This means that the actual resistance value could be 100 ohm plus 5 percent or 100 ohm minus 5 percent. Five percent of 100 ohm is 5 ohm. Therefore, this resistor marked with a 100-ohm value could in reality be between 100 ohm plus 5 ohm or 100 ohm minus 5 ohm; that is, between 105 ohm and 95 ohm.

The resistor should be oriented so that the gold or silver band is at the right-hand side and the resistance value is then read from left to right. Instead of the gold (5 percent) or silver (10 percent) band, resistors can also have no color to represent a 20-percent tolerance.

Another feature of the resistor is its power rating. Typically, a small 1/4-watt-size resistor is most commonly used, especially for the projects described in this book. The physical size of the resistor gives you an indication of its power rating. If resistors are bought several to a packet, then the power rating will be marked on the packet. The next larger power rating, the 1/2-watt resistor, can be used as a substitute for the 1/4-watt resistor but is physically larger and, therefore, takes up more board space. This is a waste because the smaller, more compact 1/4-watt resistor is used in most cases.

One of the common uses for a resistor is to limit the current flowing through a circuit. A low-value resistor allows more current to flow and a high-value resistor allows less current to flow. A good example of this property would be the case of a battery supplying current to a flashlight bulb. By placing a resistor between the battery and lightbulb, the current would be reduced and the light would be dimmed.

A wide range of resistors are available between 1 ohm and 1 Mohm. Practically, however, I find that the following range is comfortably smaller and hence more manageable. You might want to consider using this as a starter range when buying components.

Resistors	Color Code
10 ohm	Brown, Black, Black
100 ohm	Brown, Black, Brown
1 kohm	Red, Black, Red
2.7 kohm	Red, Violet, Red
4.7 kohm	Yellow, Violet, Red
10 kohm	Brown, Black, Orange
27 kohm	Red, Violet, Orange
47 kohm	Yellow, Violet, Orange
100 kohm	Brown, Black, Yellow
270 kohm	Red, Violet, Yellow
470 kohm	Yellow, Violet, Yellow
1 Mohm	Brown, Black, Green

Over the years, these 12 values have covered more than 90 percent of my project resistor needs, all in the 1/4-watt power rating value.

Resistors are robust and unlikely to be damaged. They have the added advantage over other electronic components as being the simplest to use in electronic circuits. When bending resistor leads (for soldering purposes as explained later), always make sure that the wire is not bent flush to the resistor body. Leave a small gap (about 1/16 inch is fine) so the resistor body does not fracture with the bend.

Resistors can be increased in value by the simple method of connecting them in a series, that is to say, one lead of the first resistor is connected to the second resistor. The increased value is measured across the two free ends. For example, two resistors with individual values of 10 kohm would add up to 20 kohm when connected in a series. This is a useful tip when you need a specific resistor value that is larger than you might have on hand. The final resistor value can be calculated from

$$Rtotal = R1 + R2$$

Another way of connecting resistors is in parallel. In this case, one end of the first resistor goes to one end of the second resistor, and the two free ends are also connected together. The total resistance this time is a little more complicated to calculate and is given by

$$1/Rtotal = 1/R1 + 1/R2$$

For example, if the two resistors were equal in value, say 10 kohm, then the total parallel resistance would be 5 kohm. If the two resistors were unequal in value, then the total parallel combination is always less the smaller of the two resistors. For example, a 4.7-kohm and 10-kohm resistor in parallel results in a 3.2-kohm resistor. This is another tip for creating a smaller resistor value than you might have.

CHAPTER **3**

Potentiometers

The potentiometer is a specialized form of the resistor in that it is a device that provides a continuously variable resistance. The potentiometer has three terminals. The two outer terminals are connected to either end of a circular section of resistive material. If this were a 10-kohm potentiometer, then the resistance measured across the two outer terminals would be 10 kohm. The novelty, however, is with the center terminal. This is connected to a moving track, called a *wiper*, that is in physical contact with the circular resistance material. By rotating the wiper, the resistance can thus be varied. The rotation is controlled by an external shaft onto which is mounted a knob. The entire mechanism is sealed inside a metal case for protection.

Typical values of potentiometers range from 10 kohm to 100 kohm to 1 Mohm. A common example of the use of a potentiometer is the volume control on a radio. Because these devices have three terminals, potentiometers are also called *potential dividers*. A potential divider is built up from two series resistors (R1 and R2). One end of the series resistor network is connected to ground, and the other end is connected to the signal input. In the simplest case, let the signal input be a dc voltage, that is, a battery. The signal output is then taken from the junction of the two resistors and ground. The value of the output voltage is given by

$$Vout = Vin \times (R1/R1 + R2)$$

If Vin = 9 volts and R1 = 50 kohm and R2 = 50 kohm, then

$$Vout = 9\,V/(50\,k/100\,k) = 9\,V/2 = 4.5\,V$$

What we have done is to reduce or divide the input voltage, hence the term *potential divider*. So, if you want to split a voltage equally in half, use an equal-value resistor divider network. What if you wanted a different value of output voltage? This can be accomplished by changing the value of resistor R1.

Of course, a much easier way to have a range of output voltages is to use the potentiometer instead of two separate resistors. Now, if you connect one of the outer terminals of the potentiometer to the negative side of a 9-volt battery and the other outer terminal to the positive side of a 9-volt battery, you can get a continuous range of output voltage by taking the output from the center terminal and negative 9-volt battery connection. By rotating the potentiometer shaft, you can vary the voltage anywhere from 0 volts to 9 volts. The voltage should increase as the shaft is rotated clockwise. If the voltage increases when you turn the shaft counterclockwise, just reverse the two outer terminals. The ground connection always goes to the negative supply voltage.

To get a variable resistance, you can use just the center terminal and either of the remaining two terminals. Which one you use depends on how you want the resistance to be controlled. Using one connection method (center terminal and outer terminal #1), the resistance will increase as the shaft is rotated clockwise. Reverse the connection method (center terminal and outer terminal #2) to get the opposite effect. Some circuit applications require either type, but the most common is probably to have the resistance increase as the shaft is rotated clockwise.

Potentiometers can be obtained with either a linear or audio taper, meaning that the value of resistance corresponding to the angle of shaft rotation will be different. For audio applications, the audio taper is preferred, but generally for circuit projects, such as controlling amplifier gain, the linear type is used.

CHAPTER **4**

Multimeters

An excellent way to become familiar with the resistor color codes is to verify the resistor values with a multimeter. A multimeter is an instrument used to measure voltage, current, and resistance. The most common functions you will use for hobby projects are dc voltage, dc current, and resistance. Multimeters come in two types: (1) the older but still useful analog form where the display is with a calibrated meter scale and a pointer, and (2) the modern type that features a digital readout with a number display. The difference between the two is comparable to that between the older analog watch face and the newer digital watch.

If you already own a multimeter and have a few resistors in hand, then let's put them to good use. Set the multimeter to the resistance range. With an analog meter, zero the reading with the probes shorted. The digital meter doesn't need zeroing. Place the meter leads across the resistor, making sure that your fingers are not touching the resistor leads, and read the ohm value. This is a good way to check if your color code readings were correct.

Always switch the analog meter to the off position after you've finished measuring resistors. The internal battery is connected into the circuit during resistance measurements and will run down if left connected (with an external resistance load). It is quite all right, though, to leave the analog meter on the voltage or current range because the internal battery is not used.

Digital meters, on the other hand, should always be switched off when finished with, regardless of which measurement is being made, because the internal electronic circuitry is used for all measurement functions.

Whether you're using an analog or a digital meter, the multimeter leads (generally the positive red lead) will have to be physically swapped to a different socket when making current measurements. Your multimeter's operating manual will give you the precise instructions on which terminals to use. Some multimeters also have an internal fuse to protect the unit in case the current being measured is excessive; this particularly applies to digital multimeters. Should this happen, and your meter appears not to respond to current measurements, check the internal fuse. Excessive voltage overloads cannot damage (within reason) digital meters because the unit will either automatically switch to the next highest range or indicate an overload.

Analog meters, on the other hand, should be protected from the application of an excessive voltage to the instrument. The meter needle will physically hit the end stop of the scale and could be damaged, depending on how high an excessive voltage is applied. With an analog meter, always set the unit to the highest range when measuring an unknown voltage and then gradually reduce the range until the meter starts to read correctly.

Analog and digital meters each have their own individual benefits. I own both types (simple, inexpensive models) and use them frequently. Analog meters are great for determining how a voltage is varying by watching the meter needle track a varying input voltage. Digital meters, on the other hand, are generally much easier to read because the voltage is already displayed as a numeral. Analog meters are not protected against voltage polarity reversal. In other words, if the meter leads do not match the polarity of the voltage source, then it is highly likely that severe damage will result. Digital meters, on the other hand, will merely show a negative voltage to indicate that the polarity has been reversed.

Regardless of whether you decide to get an analog or a digital meter, it is the essential piece of test equipment that you must own. Buy the best type you can afford at an electronics store, but avoid those found in hardware stores, which are used mainly for electrical installation work.

CHAPTER **5**

Ohm's Law

Three fundamental measurements in electronics form the basis of most of the simple calculations you will need: (1) current, (2) voltage, and (3) resistance. These three units of measurement are tied together by Ohm's law, which states that when a voltage is applied to a circuit, the resultant current that flows is inversely proportional to the total resistance in the circuit. This relationship is given by

Current flowing through circuit = Voltage applied to circuit ÷ Resistance of circuit

or

$$I = V/R$$

where voltage is measured in volts, current is measured in amperes (amps), and resistance is measured in ohm.

Ohm's law can also be rearranged to give two other equations. First,

Voltage applied to circuit = Current flowing through circuit × Resistance of circuit

or

$$V = I \times R$$

Second, concerning the measurement of resistance,

Resistance of circuit = Voltage applied to circuit ÷ Current flowing through circuit

or

$$R = V/I$$

These are three extremely useful ways of expressing Ohm's law. You will find that the most commonly used equation is the first one we considered: $I = V/R$.

In most of the instances when you will be applying Ohm's law, the voltage is generally known (usually 9 volts, as with the circuit projects included in this book). The resistance will also be known (as given by the value of the resistor in the schematic). Therefore, we most commonly need to find the current flowing through the resistor.

For example, assume that you have a 9-volt battery and a resistor of 10 kohm. When the resistor is connected across the battery, current will be flowing through the resistor. The amount of current is found by

$$I = V/R$$

or

$$9\,V/10\,k = 0.9 \text{ milliamps}$$

In another case, assume that we had a resistance of 10 kohm and required a current of 0.9 milliamps to flow. What voltage would be needed to produce this current? Again, from Ohm's law, the required voltage is given by

$$V = I \times R = 0.9 \text{ milliamps} \times 10\,k = 9\,V$$

As a final example of the application of Ohm's law, assume that we have a voltage of 9 volts and require a current of 0.9 milliamps to flow through a circuit. What value of resistor is now needed to limit the current to this value? Ohm's law says

$$R = V/I = 9\,V/0.9 \text{ milliamps} = 10\,k$$

CHAPTER **6**

Light-Emitting Diodes

Light-emitting diodes (LEDs) are simple, two-terminal, modern-day equivalents of a regular flashlight bulb. There are two important advantages to using an LED as an indicator lamp rather than the filament bulb you would find in a flashlight. First, the LED is considerably more robust than the filament bulb, and hence can be subjected to much mechanical abuse. Second, the LED draws significantly less current than the filament bulb, thus making it a perfect complement to modern-day electronic project designs.

The LED is always used as a status indicator to show the presence of a voltage, most commonly as an indicator for the power supply circuit. The LED is polarity sensitive, which is to say it must be connected the correct way around to the power supply voltage, such as a 9-volt battery; however, no damage is done if the polarity is reversed. Again, this is a robust device. Incidentally, when the LED is on in the so-called forward bias mode, current is able to flow through this essentially solid-state lamp. In the reversed connection, or the reverse bias mode, current cannot flow through the device, and hence the LED is off.

On one common type of LED, the negative terminal is identified with a flat edge on the side of the component, but if this is not immediately visible or there happens to be a different marking protocol, there is no cause for alarm. It is actually much easier to find the correct connection as follows: You will need a 9-volt battery, a 1-kohm resistor (actually the value here is not critical because anything between 1 kohm and 10 kohm will work), and the LED. Connect one end of the resistor to either end of the LED. Now connect the other free end of the LED to one terminal of the battery and the other free resistor end to the other battery terminal. The LED will light up if the polarity connections have been made correctly. If the LED does not light up, simply reverse the connections to the battery. Now the LED will light up and there has been no damage to the device. Whichever way causes the LED to light, take note of the polarity connections of the LED and the physical markings on the device.

Want to experiment a stage further? Change the resistor value to 10 kohm and now notice how much dimmer the LED is, because there is less current flowing through the LED. That current flow can easily be measured by connecting a multimeter (set to the dc current range) in series with the battery's positive supply line.

As we learned in Chapter 5, the current can be calculated using Ohm's law,

$$\text{Current} = \text{Voltage}/\text{Resistance}$$

or, using the 1-kohm resistor,

$$I = V/R = 9\,V/1\,k = 0.009A \text{ or } 9 \text{ milliamps}$$

In the case of the higher-value resistor, we can similarly calculate the current flow. Thus using the 10-kohm resistor, we get

$$I = V/R = 9\,V/10\,k = 0.0009A \text{ or } 0.9 \text{ milliamps}$$

Where an LED is used as a power on indicator, take note of the additional current drawn by the LED when measuring the total circuit current consumed. For battery-powered circuits, as is the case with all of the construction projects in this book, the more current used, the quicker the battery will run down. For the absolute maximum battery lifetime, use the highest-value resistor (typically 10 kohm) that will still give a usable light output, which is a contribution of just under 1 milliamp. This is quite a good saving of current.

CHAPTER **7**

Switches

The basic purpose of a switch is to perform a simple on/off function. Most commonly, this task is accomplished with a toggle lever, but it can also be done with a slide-action control. Look at any piece of electronic equipment—an amplifier or a radio, for instance—and you will see examples of these two types.

Switch types are identified by *poles* and *throws*. When you go to an electronic parts store to buy a switch, you must first know the number of throws and poles needed. To understand these terms, we need to be familiar with the circuits that are connected to switches. First, if you look at electronic circuit schematics, you will see that switches are by convention connected in the positive supply line only. Switches are never situated in the common ground line.

Second, there is always by definition an input side and an output side to a switch. Conventionally, the side of the switch receiving the power (from the battery) is termed the *input* side. For example, when we connect a battery to an LED, the battery is referred to as the *input* and the LED as the *output*. Therefore, we will need a switch with two terminals—one terminal to which the battery input is connected and one terminal to which the output is connected. In one position of the switch toggle, the terminals are open circuit; in the other position, they are short-circuited.

This condition is easily verified. Take a regular on/off switch that has two terminals and connect the two leads from a multimeter (set to the resistance range) across the switch terminals. In one switch toggle position, the meter will read 0 ohm, or a short-circuit, and in the other switch toggle position the meter will read infinity, or open circuit.

This type of switch is called a single pole, single throw (SPST). *Pole* refers to the number of terminals to which an input can be connected (in this case it is one), and *throw* refers to the number of terminals to which the output can be connected (which also in this case is one).

A second common type of switch is called a single pole, double throw (SPDT). The difference between the SPST and the SPDT is that the former has two terminals, whereas the latter has three. When you come across the three-terminal SPDT type of switch, the input will always go to the center terminal (the terminals are arranged all in a row). The output can go to either of the remaining outer two terminals.

The SPDT type of switch is designed, however, to be more than a mere on/off switch. If you want to switch a single input to two outputs, then the SPDT is the type of switch you would use. All three terminals are used. The input always goes to the center terminal, and the two outputs go to the remaining two terminals.

Once more, let's check this out using an SPDT switch and a multimeter set to measure resistance. Connect one of the meter leads to the center switch terminal and place the other meter lead to either of the remaining switch terminals. Note the meter reading. Toggle the switch to get a 0 ohm or short-circuit reading. Now flip the toggle in the other direction. The meter will read infinity or open circuit. Without changing the toggle position, place the meter lead (that is not connected to the center terminal) to the remaining switch terminal. The meter now reads 0 ohm or a short-circuit. You have shown that the input connected to the center terminal can be alternately switched to either of the two outer terminals (or outputs). Take note that a three-terminal SPDT switch can also be used as a simple on/off switch by merely using the center and one of the other terminals.

CHAPTER **8**

Capacitors

Next to the resistor, the second most common type of electronic component you will come across is the capacitor. A special property of the capacitor is its ability to block dc voltage and pass ac voltage. It is commonly used for coupling signals into and out of audio amplifiers and is also used across the power supply lines to smooth out any voltage fluctuations.

The unit of capacitance is the farad, but because this is rather large, it is broken down into smaller units of measurement. There is the microfarad (µF), which is one-millionth of a farad, and the picofarad (pF), which is one-trillionth of a farad. At this point, it is necessary to go into the numbering system in a little more detail. Small-value capacitors, typically between a few picofarads and about 1,000 picofarads in value, use the picofarad unit abbreviation—pF.

Larger capacitor values, from about 1,000 picofarads and upward, are generally given the microfarad suffix—µF. From the definitions of the picofarad and the microfarad, we can derive the relationship between the two: There are one million picofarads to a microfarad. Armed with this information, we can then see that 1,000 pF can also be expressed as 0.001 µF. Thus,

    1,000 picofarads (pF) = 0.001 microfarads (µF)
    10,000 picofarads (pF) = 0.01 microfarads (µF)
    100,000 picofarads (pF) = 0.1 microfarads (µF)

The higher the number of picofarads, the more unwieldy that unit of measurement becomes and the easier it is to use microfarads instead. For example, instead of asking for a 100,000-picofarads capacitor, it is much easier to ask for a 0.1-microfarad capacitor. You might also come across another capacitor size, the nanofarad (nF), which is between the picofarad and microfarad in size. One nanofarad is 1,000 millionths of a farad; however, I suggest you ignore the nanofarad and just use the picofarad and microfarad because you can cover all of the necessary capacitor values with these two prefixes.

The smaller-value capacitors are nonpolarized, which means it doesn't matter how you connect the leads. Capacitors with larger values, however, are polarized, which is to say that the leads are identified as positive and negative and must be correctly connected. A popular type of polarized capacitor is called an *electrolytic capacitor*. Typically, from about 1 µF upward, capacitors are of the electrolytic type. They are also physically larger than nonpolarized types and will have a lead emerging from both ends of the tubular-shaped capacitor or have two leads coming out of one end. Both types are commonly used for projects.

A voltage rating for capacitors defines the highest supply voltage they should be subjected to. All capacitors are two-terminal devices. Unfortunately, it is not a simple case to carry out basic tests with capacitors as it is with resistors because they are fundamentally ac rather than dc components.

CHAPTER **9**

Integrated Circuits

The components so far described all come under the category of passive components, which means that they perform only the fundamental function for which they were designed. For example, resistors provide an electrical resistance to current and no more than that. Another class of components, however, is called *active components*, which are much more versatile and which, by varying the design of the circuit, can provide a wide range of ingenious circuit functions.

The integrated circuit (IC) is an example of an active component. The IC comes in a wide variety of shapes, sizes, lead counts, and basic circuit functions. There are far too many to list here in their entirety, so we will restrict ourselves to the three types that are used for the construction projects described later in this book (LM 555, LM 741, and LM 386).

Most ICs work off a positive supply voltage and function nicely from a regular 9-volt battery. Two of the most common functions performed by ICs are amplifiers and oscillators. An amplifier takes a small, low-level signal and increases or amplifies it sufficiently to drive a speaker or headphones. An oscillator is a signal source that is commonly used to provide audible evidence of the correct functioning of a circuit, for example that of an amplifier. Integrated circuits require additional components, generally resistors and capacitors, before they can provide a useful function. Hence, simple tests cannot be done here.

Integrated circuits fall predominantly into two broad categories: analog and digital. Analog circuits are characterized by the fact that the signals being processed by the device can vary from any value between zero and the maximum allowable (generally close to the supply voltage). Digital circuits, however, have only two levels of voltage operation. The level is either off or low, which corresponds to a value that is small, or the value is on or high, which corresponds to a value that is large. The polarity requirements for both analog and digital circuits are such that the power supply rail is almost always positive.

Integrated circuits are superb devices because they incorporate a large number of discrete components that would individually take up an impractical amount of space. By way of an example, in the days before the advent of ICs, the only way to build, say, a battery-powered guitar audio amplifier, was to construct it from discrete components using transistors. This required a large number of components, took up quite a lot of space, and was somewhat difficult to set up. Now, using an audio-power IC, the number of components needed to build the same amplifier is quite small, and the amplifier takes up little space and is incredibly easy to set up.

All of the projects you find in this book use ICs to enable complete functions to be performed easily. In no way would this be possible if discrete components were to be used.

CHAPTER **10**

Tools

To get started with the construction of an electronics project, you will need a few basic tools. The most important tool is, of course, the soldering iron, so let's start there. Soldering irons come in a wide array of sizes and power ratings. I have used a small 25-watt iron with a fine tip for more than 20 years. It is small enough to handle integrated circuits and strong enough for all of the hundreds of circuits I've built over the years. Regular solder called *rosin core solder* is what you will need for soldering components to the assembly platform.

Only a few other tools are required. The wire stripper will probably be the most useful item around and is used, as the name implies, to strip away insulation from wires before you solder them. The same basic model I bought in 1965 is still going strong today! Small needle-nosed pliers are essential for holding component leads in place during soldering or for bending resistors and solid wires to shape. Good-quality miniature cutters are needed for general-purpose cutting of wire or trimming of component leads after soldering. Do not be tempted to use these cutters for cutting anything larger than hook-up wire because the cutting edges will likely be damaged.

A variety of screwdrivers with different tips (slot and cross shapes) are also needed for volume control knobs and project cases. For drilling holes in project cases, a small battery-powered drill is useful. If you can get hold of one, a reamer is a terrific tool for enlarging holes to size. Needle-nosed files allow for precision cleaning up of holes. An inexpensive glue gun is a neat way of securing mechanical components. Finally, a set of miniature open-ended wrenches is needed for tightening up nuts.

When using wire cutters to snip away excess component leads, take extra care with the cutting process. The wire that is snipped off can fly away with great speed and force, so it is essential to make sure that it does not cause injury (direct the trajectory away from your eyes); also make sure that it does not cause short-circuits by being lodged in your circuit. The best way is to direct the cut-off pieces downward onto your workbench. Dispose of these pieces as soon as possible.

As a precaution, always remove your soldering iron from the power source as soon as you have finished soldering. There is generally no visual indicator on the iron to show that the tip is still being heated, and in the interest of safety, make sure that any extraneous material is kept well away from the very hot tip.

CHAPTER **11**

Soldering

The key to a project's working the first time you switch the power on is strongly tied to the quality of your soldering. This chapter covers the most common soldering faults contributing to a dead circuit.

Most notorious are cold solder joints, which are solder joints that visually appear acceptable but where, in fact, the solder is merely lying on top of the joints. How does this happen? Usually, insufficient heat from the soldering iron tip has been applied to the joint prior to the solder being melted. All that has happened, therefore, is that the solder has melted and sat on top of the joint. Any mechanical stress to the joint will soon show that the joint was never made in the first place. It is vital, therefore, to adequately preheat the junction between the two components to be soldered together before the solder is applied. Practice is the only way to gain proficiency. Start by soldering onto a bare piece of track even without components. Of course, it goes without saying that you must have the proper tools for soldering.

A 25-watt soldering iron with a miniature tip in good condition is essential. Clean the tip with a moist sponge (you can purchase this at the same time you buy a soldering iron) and slightly coat with a layer of solder before soldering. Place the tip of the iron in contact with the components and press firmly but not too hard. Wait a few seconds for the heat from the iron to transfer to the components, then apply a small amount of solder to the preheated junction. The solder will melt. Remove the solder and then remove the iron. Keep the connection secure until the solder has cooled (about 5 seconds). A smooth, shiny solder joint is the sign of a good connection.

Insufficient solder is another error. There should be enough solder to cover the junction of the two components to be soldered, but not so much that the junction cannot be seen. Bridged solder between closely spaced tracks is yet another common fault, and the best way to spot this error is to use a magnifier to check your soldering. To remove the bridged solder, place a heated iron quickly on the excess solder. If an excessive amount remains, then use either a solder sucker, which will suck up the heated solder by vacuum action, or a solder wick, which is a specially formulated braided wire for sucking up melted solder. Either tool works fine.

The best advice regarding soldering is to practice, and practice until you become proficient. Even soldering onto a blank circuit board without any components is an excellent way to develop your skill. Aim for the minimum amount of solder necessary and the shortest duration of heat application, and you can't go wrong.

CHAPTER **12**

Basic Electronics Theory

When you construct an electronics project, the end result is that the circuit performs a particular function. For example, when you are finished building an amplifier, the circuit will typically amplify a small, low-level signal to a level sufficient to drive, say, a loudspeaker.

Several commonly used electronic building blocks can be linked together to get a somewhat different circuit design. For example, let's continue to consider the amplifier. There are typically two distinct types of audio amplifiers. The preamplifier first amplifies a low-level signal, for example the signal coming from a record player, and might also add the bass and treble controls. Then there is the power amplifier, which supplies the driving power to the loudspeaker or headphones. The preamplifier and power amplifier can be constructed separately or combined. In general, amplifiers will, therefore, always have an input side, to which an external low-level signal is fed, and an output side, from which a high-level signal is taken to drive another device.

A second common type of circuit is an oscillator, which generates an audible audio signal, often variable in frequency. The signal output can be low level or high level and is used for feeding into another circuit stage, again typically something like an amplifier. Oscillators do not have an input connection and have only a single output connection.

Power for any electronic circuit is always from a dc voltage source. Because the line voltage is an ac source, power supply circuits convert the high ac line voltage into a lower, usable dc voltage. For all of the simple construction projects described later in this book, however, the dc power source is a regular 9-volt battery. The current drawn by any of the circuits you'll be building is very small, and a 9-volt battery is a more than adequate power source.

The dc power supply always has two voltage polarities, a positive terminal and a negative terminal, just as all batteries are marked with a positive and negative terminal. Take a look at a 9-volt battery and you'll see the positive terminal marked.

Electronic circuits thus have a positive voltage side and a negative voltage side. There is always also a common ground connection, and most (although not all) circuits have the negative voltage end connected to ground. All the circuits described later have the negative voltage end connected to ground. That means the positive supply voltage is always referred to as the *supply voltage*.

A typical electronic circuit such as an amplifier can thus be simply represented as a black box or block diagram with four terminals. The upper terminal is the positive supply line, and the lower terminal is the ground or negative supply line. To the left is the input line, and to the right is the output line.

CHAPTER **13**

Reading Schematics

A circuit schematic is a drawing of the electrical connections needed to make the circuit function. Such a diagram uses a series of standard electronic symbols to represent the various components, such as resistors and capacitors, that are used in the project. All of the electrical information is given in the schematic. Translating the schematic into the actual physical connection of components requires knowledge of the components themselves and how they should be connected together. In the circuit schematics associated with the construction projects described later, I have made sure that these have been uniformly drawn so that they can be easily read by the beginning hobbyist.

You should know some simple guidelines for reading a schematic. Start by looking at the top half of the schematic for the common rail running across from left to right. That rail will most commonly be the positive power line rail. At the bottom of the schematic, also running from left to right, will be the ground rail, which is generally connected to the negative power supply line. If this schematic has an input (such as an amplifier), then an input line will be at the left-hand side of the schematic. Finally, there is almost always some form of output, which is the line situated on the right-hand side of the schematic.

So, now you know that there are typically four basic sides to a schematic and generally what these lines should correspond to. Let's add some detail. Along the upper top rail, which we now know to be the positive power rail, there will be several standard components. First of all, the power switch is situated in the positive power rail. With all of the circuits shown here, the switch is always positioned along the right-hand side. Incidentally, the battery, which is of course needed to power these circuits, is always positioned at the extreme right-hand edge. These conventions are purely my own, but following them consistently makes reading schematics much easier.

Next to the power switch there is sometimes a light-emitting diode (LED) indicator light, which is always positioned on the side of the switch that is farthest from the battery. If you have an LED, then there must always be a resistor to limit the current. Next, as an aid for good circuit stability, two capacitors, an electrolytic and a disc ceramic, are positioned on the same side as the LED and run directly across the positive and negative rails. Thus, you now know that in general a circuit schematic will almost always contain a battery, switch, LED, resistor, and two capacitors. Consult the 10 projects diagrammed later in Chapter 17 and you will see this is true. These are standard circuit components, and regardless of the type of circuit, whether it is for an amplifier or an oscillator or for something else, you will find these components.

Another two components can also be added to our list. If the circuit schematic is for a simple amplifier, it is likely that the amplifier is an ac type, in which case there will be a capacitor at the input side and another capacitor at the output side. So, here are two more easily recognized components.

Finally, simple battery-powered circuits using op-amps such as the LM 741 require that the integrated circuit (IC) is fed with a split power supply, that is, a dual positive and negative supply. One way to do this is to use two batteries, but this is a cumbersome solution where it is not critically needed. Instead, a common arrangement is to use three components to generate the split supply. Two equal-value resistors are taken from the positive and negative rails, and the junction is tied together. If we have a 9-volt battery supply, then the midpoint will be half the voltage, or 4.5 volts. A capacitor is the third component we can use to provide a split power supply, as it is placed across the junction and ground to provide smoothing.

In summary, we have accounted for a total of 11 components so far that are almost always present in basic electronics projects—if you were building an ac amplifier, for example. That is a significant number of components that can be classified as common components, thus making the task of interpreting schematics for the beginner much easier. When you come across a new circuit, start looking for these common components. After you've accounted for them, the circuit will look less daunting.

Circuits invariably have a few components that are needed purely for the IC to operate correctly as opposed to contributing to the circuit's functionality. The 555 timer, for example, always has pin #5 connected to ground via a capacitor. So, there's at least one component you can immediately forget about when trying to figure out what the circuit does.

All of the circuit projects in this book are logically designed so that the commonality of components is clearly seen. In addition, a suggested drawing layout of the components on the schematic is drawn consistently from project to project, again making component recognition simple.

CHAPTER **14**

Assembly Techniques

Provided you have the right tools in place, components, and a circuit schematic in front of you, we can make a start with some general guidelines on how to maximize the chances of your projects working perfectly the first time. Let's start with some basic techniques.

All circuit projects, regardless of type, require wires to be soldered into the circuit, acting either as a bridging link on the circuit board or as a means to get connections onto the circuit. For example, if you are building an amplifier and want to power a speaker situated away from the circuit board, wires (two in this case) would have to join your amplifier to the speaker. Basically two types of wires are commonly used when making solder connections: (1) solid hook-up wire and (2) stranded hook-up wire, both of which come in many different sizes (or gauges). The key difference between the two types is that the solid type is a single, solid length of wire, whereas the stranded type consists of several fine strands of wire. With both types, the wire is covered with insulation.

If working with solid wire, use wire strippers to strip back and remove the insulation from a short length of wire, typically about 1/4-inch long. Make sure that when you strip off the insulation, the wire itself is not nicked. If this happens, cut off the end with wire cutters, increase the gap of the wire strippers, and start again. A nicked wire is weakened and prone to break off under stress. The exposed wire is inserted into the assembly board and then ready for soldering.

For the stranded wire, remove the insulation as before, but this time twist the loose strands together before using them. Leaving the strands untwisted causes them to splay apart when solder is applied. For circuit board connections, either solid or stranded wire can be used. I use both types interchangeably, depending on the application. For making connections externally, stranded wire is recommended because the stresses arising from moving components could lead to fractures if solid wire were used. In selecting the suitable wire size, use the diameter that will comfortably fit through a regular circuit board. If you are unsure about exact sizes, always go a little smaller than needed.

We've covered wires, so now what's next? As discussed in Chapter 11, good soldering technique is vital if your circuit is to function reliably. The key to good soldering results is to use the correct size of soldering iron, have a clean tip, and use good soldering practices. An excellent size of soldering iron is the 25-watt iron with a fine tip. Too small an iron means there is insufficient power to heat up the components. Too large and there is the danger of overheating the components. I have always used a 25-watt size and have even found that the 15-watt and 35-watt sizes are unsuitable.

Good soldering practice begins as follows. Have a moist sponge handy next to your soldering iron. Once the iron has reached the right temperature, briefly clean the tip with the moist sponge, and then apply a small amount of solder onto the tip. If the iron has warmed up properly, the solder will flow quickly and smoothly, leaving a shiny deposit on the tip. Wipe off the excess solder on the sponge, if necessary. The iron is now ready to use. Do this just before you are ready to solder. Bring the tip to the junction between the two components to be joined. Keep the tip firmly in place for a few seconds, thus heating up the junction. Now, without moving the tip, apply solder to the heated junction. The solder will flow quickly and smoothly. Remove the solder and then remove the tip. Hold the components firmly in place while the solder cools.

A good solder joint is shiny, smooth, and made with the correct amount of solder, which means you should still be able to discern the outline of the components underneath. If all you see is a huge mountain of solder, then you've used too much. Too little solder, on the other hand, means that you can see bits of the junction between the components not covered with solder. As with all skills, practice is the only way to gain proficiency.

We already discussed in Chapter 11 a type of solder error called a cold solder joint. Because it is particularly troublesome, it's worth reconsidering it here. This solder joint looks acceptable but is in fact only resting on top of the junction. This type of fault occurs when the junction has not been sufficiently preheated before applying the solder. As a consequence, any later excessive force or stress will disturb the cold solder joint, leading to an open circuit.

Another error to be cautious of is solder splashes or bridges between tracks. This mistake is easily made, especially when adjacent tracks are close. The solution is to be careful about how and where the solder is applied and to inspect each joint carefully with a magnifier. Minor excess solder is easily removed by carefully reheating the unnecessary solder and removing it with the iron tip. After each solder operation, clean the tip by wiping it on the moist sponge, if necessary. Every now and again, re-tin the tip as needed.

When inserting components such as resistors into the circuit board prior to soldering, bend the leads to size and insert them into the board, but without cutting to size. Only after the leads have been soldered should you trim the leads to size (on the solder side of the board). The component should be close to the board but not absolutely flush. I like to leave a little space underneath the component so that, if need be, I can gain leverage underneath it for removal. The actual positioning, though, is not critical and will not affect the functioning of the circuit. Generally, to have a nice compact unit, components should lie fairly close to the board.

If you find that a component has been soldered in the wrong position, there are two ways to remove it. The first uses a tool called a solder sucker, which is a spring-loaded vacuum sucker that is held against the heated solder. Depressing the release button releases the spring, thus sucking up the molten solder into the housing. The second way is to use solder braid to soak up the molten solder again. Both methods are good.

CHAPTER **15**

Handling Components

Resistors must be bent into shape before insertion into the circuit board. The leads always emerge from the body of the resistor—one from each end—and are commonly bent at right angles so they can be inserted into the circuit board. The leads should be a short distance (a few millimeters) away from the end of the resistor body, otherwise there is the possibility of the resistor material fracturing.

There are two ways to bend resistors. The first is to use a pair of needle-nosed pliers to gently grip the leads and bend them by hand. The pliers not only act as a convenient gripping tool, giving you a nice straight edge, but also and more important, the pliers buffer the lead just enough from the resistor body. The second way to adapt resistors for insertion into the circuit board is to bend just one of the leads along the length of the resistor body and to leave the other lead intact, so that you end up with both leads facing the same direction. Again use the pliers to maintain the small spacing from the resistor body end.

Why the two different methods? The first type of bend is used where there is plenty of space and the resistor can be placed parallel with the board; however, where space is limited and you need to place a resistor between two adjacent pins on a circuit board, the second bend method is used.

Capacitors (disc ceramic types) rarely need bending because the leads already emerge from the same side of the capacitor body. Take care, however, that the leads are not stressed, or the area of the capacitor from which the leads emerge may be cracked. This could happen if the capacitor is pushed flush with the circuit board, so keep the capacitor a short distance away from the board. Where you need to open the lead spacing by bending, use the same needle-nosed pliers to grip the end of the lead nearest the body. By doing so, that part of the capacitor is protected against any strain.

Electrolytic capacitors can be purchased with leads coming from either end of the body (these are called *axial lead devices*) or with leads coming from the same end of the capacitor (these are called *radial lead devices*). I use only the radial lead devices to reduce the amount of components I carry.

Determining which type of capacitor should be used is logical. Where surface height is tightly restricted, for example in a pager, the axial type of device would be used. Where plenty of headroom is available, but board space is limited, the radial device should be used because it conserves board area. For simple hobby projects, board space is more valuable and height restrictions are never an issue, so radial devices are used here.

Integrated circuits (ICs) are not handled in the same way as resistors and capacitors but often are inserted into sockets. All you need to do here is to squeeze the leads slightly inward to fit the socket. The socket itself is soldered directly to the circuit board and has no special needs.

CHAPTER **16**

The SINGMIN PCB Circuit Assembly Board

The layout designs for the projects in the next chapter are for a dedicated assembly board—the SINGMIN PCB—which I designed and manufactured in response to a lack of suitable assembly boards available at that time. Since then, however, with continual advances being made in the availability of electronic components for the hobbyist, there are likely more efficient assembly boards available today. My original layouts can be used as a template for laying out your assembly board of choice. Thus, the parts lists for each project refer to a general-purpose assembly board rather than the SINGMIN PCB board. The circuit projects described in this book give full schematic details and a suggested component layout scheme for total completeness. Armed with those two diagrams, you should have little difficulty getting the circuits to work the first time.

The SINGMIN PCB emerged out of the need to find a better solution to the current methods for prototyping simple projects. The top view of the SINGMIN PCB is shown in Figure 16-1. The way it is shown in the figure, with the legend "SINGMIN PCB-1" at the top of the board, is the way you would correctly orient it for your use. The size of the board is 4 inches by 4.5 inches, although it may be split or quartered into separate, self-contained boards. Separating the SINGMIN PCB into individual sections is done within seconds by merely grasping the board and snapping along the guidelines running horizontally and vertically through the center of the board. For a neater finish, you can smooth down the edges of the board after splitting them apart, but this step is not critical.

Not shown in Figure 16-1 is an extra shorting track running along the center of the board (from left to right) that can be used as an optional grounding link under the integrated circuit that would straddle this link. It is not used for any of the projects described here, and hence is left off the figures for clarity. It does serve, though, as a nice identifying check that your integrated circuit (IC) is positioned correctly, in that this shorting track must lie under the IC (running left to right).

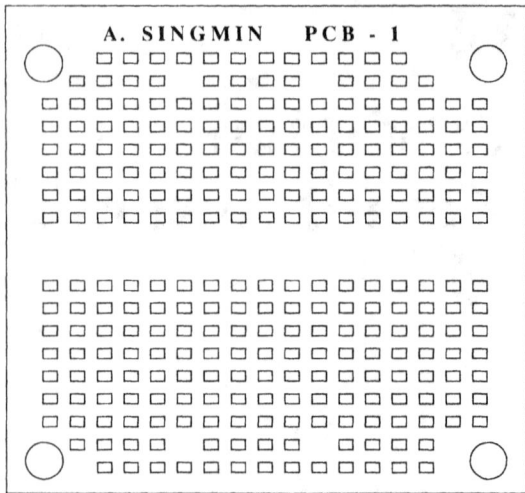

Figure 16-1 Top view of the SINGMIN PCB universal circuit assembly board

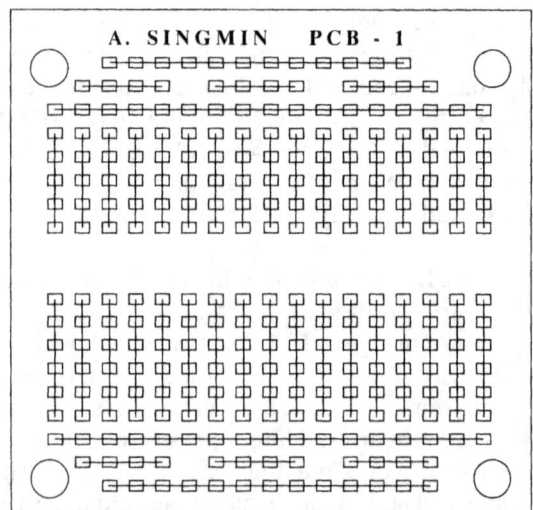

Figure 16-2 Superimposed view of underside tracking

The underside of the SINGMIN PCB interconnects various rows and columns together. The markings on the top of the board show you the routing of the interconnections. This arrangement is shown in Figure 16-2. Notice that there are two distinct and separate regions on the board. The central area consists of columns that are connected from top to bottom. The upper and lower sections consist of rows connected from left to right. This particular arrangement has been developed to complement the current method of using ICs in

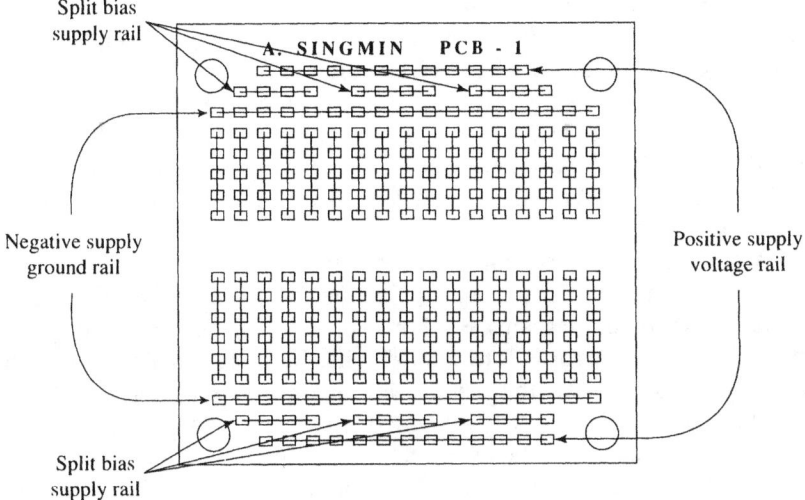

Figure 16-3 Identification of key voltage supply rails

hobby projects, thus obviating the need to provide a split supply voltage when using the typical single 9-volt battery.

Integrated circuits also often require connections to the positive and negative supply from either the upper half or lower half of the IC (you'll see this later in the projects chapter). To accommodate these requirements, the arrangement shown in Figure 16-3 identifies these particular areas for connection. Linking is always done on the top surface (so that the interconnection is always visible), typically so that the positive supply rail is available both at the top and bottom of the board, and similarly so for the negative rail. Because of the generous amount of solder points available, there is plenty of room for the simple projects later described in this book. Hence, there is a lot of flexibility as to where components can be placed. Nothing could be simpler and more satisfying for the beginner than having an easy-to-use board that eliminates the tedium of trying to fabricate a makeshift assembly medium.

The use of an IC socket is well worth the extra effort because the benefits to be gained are immense. Any errors in locating the IC can be easily rectified merely by unplugging the device and reinserting it the correct way. In the unlikely event of an actual IC failure, replacement could not be simpler.

It is highly recommended that you become familiar with handling and placing the electronic components before attempting any soldering. Components will not only vary in size but there will also be considerable differences in the way components are bent to shape and actually inserted. Always start with the IC socket because this gives you a starting frame of reference from which the rest of the components can be added. Insert the IC socket and bend the leads back to secure. Because of the simplicity of the circuit projects described here and the large board area available for placement of components, the IC socket need not be rigidly located as per the described layout. By referring to the circuit schematic and the suggested layout, you can see where the electrical connections have to be made. Once you adhere to that requirement, the mechanical placement can be anywhere that is convenient. To gain initial familiarity with circuit construction, space out the components as much as possible.

After all of the on-board components have been inserted, check thoroughly from both the top and underside for correct alignment with the schematic and layout diagrams. Be thorough at this stage because it is easy to correct a placement error before soldering has been started. After the components have been soldered, the rectification of a mistake becomes a little more difficult because some dexterity is needed to desolder components. But, in the end, learning to correct mistakes just takes practice, as with any task.

A further suggestion is to insert one component at a time and then solder because there should always be a careful check made of the integrity of each solder joint. This is easier to do if the inspection is done as you go along, after each solder joint is made, rather than at the end of the soldering stage. The time spent at the inspection stage is well worth the satisfaction of seeing the circuit work the first time the power is applied.

CHAPTER **17**

Construction Details for 10 Simple Projects

Project #1: Fixed Low-Frequency LED Flasher

It is highly recommended that Project #1 be built as a learning exercise because all of the detailed assembly and handling instructions you will need for the later projects are only given with this project. Projects #2 through #12 assume that you have already acquired some project construction experience.

Project #1 is one of the simplest circuits to begin with. This circuit, when completed, causes a light-emitting diode (LED) to flash or pulse alternately on and off at a very slow rate, that is, at a low frequency. It could be used as a dummy car alarm indicator. You sometimes see these tiny pulsating lights in expensive cars protected by sophisticated car security systems, warning you that the car alarm is armed and ready. For this design, the LED flashes on for a brief period and stays off for a relatively longer period. Because additional current is consumed from the battery every time the LED comes on, this long rest period conserves battery power. The brightness of the indicator LED is also limited, again in order to preserve battery power.

Circuit Description

The 555 timer integrated circuit (IC) is configured in an oscillating mode that results in a continuous train of pulses being generated. The frequency-determining components have been chosen to give a very slow pulse rate. Two resistors and a capacitor determine the flashing rate of the LED. Power is supplied from a 9-volt battery.

Parts List

Semiconductor
IC1: LM 555 timer

Resistors
R1 = 100 kohm
R2 = 10 kohm
R3 = 1 kohm

Capacitors
C1 = 10 μF
C2 = 0.01 μF
C3 = 0.1 μF
C4 = 100 μF

Additional Parts and Materials
LED1: Light-emitting diode
S1: Miniature SPST toggle switch
B1: 9-volt battery
9-volt battery snap
8-pin IC socket
General purpose circuit assembly board
Hook-up wire (solid and stranded)

Pin Connections (see Figure 17-1)

Pin #1

This is always connected to ground. Because all of the circuits described here use the positive voltage as the supply voltage, that means that the ground connection is the same as the negative battery terminal.

Pin #2

This is first connected to pin #6 and then is connected via capacitor C1 (10 μF) to ground. The value of the capacitor is one of the components that determine what the output frequency of the oscillator will be. A high-capacitor value results in a lower frequency, whereas a low-capacitor value raises the frequency.

Pin #3

The output frequency is taken from this pin and will go to the display LED1 to give a visual indication of the frequency. To limit the current flowing through LED1, a series resistor, R3 (1 kohm), is used. A high-resistance value conserves power but results in a dimmer light output.

Pin #4

This pin always goes to the positive supply voltage.

Pin #5

This pin is generally taken to ground via a capacitor, C2 (0.01 μF).

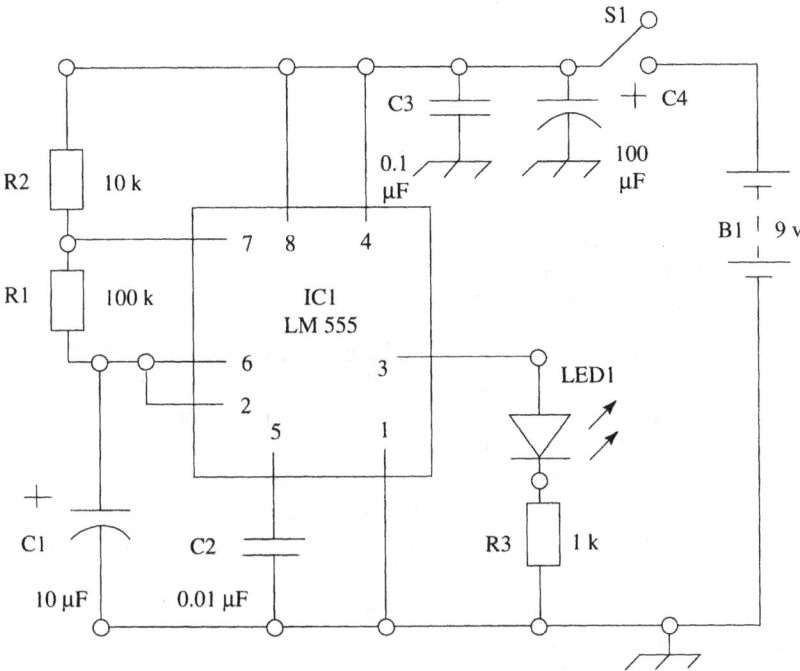

Figure 17-1 Project #1: Fixed low-frequency LED flasher

Pin #6
This pin is connected to pin #2 and is connected to pin #7 via a resistor, R1 (100 kohm).

Pin #7
This pin is joined to pin #6 via the previous resistor, R1, and is connected to the positive supply voltage via a second resistor, R2 (10 kohm).

Pin #8
This pin is always connected to the positive supply voltage. A disc ceramic capacitor, C3 (0.1 µF), and an electrolytic capacitor, C4 (100 µF), combination is connected across the positive supply line and ground. These components are used to stabilize the operation of the circuit and are found with all the projects described herein.

Component Identification

Figure 17-1 shows the electrical schematic for the fixed low-frequency LED flasher. Referring to the above list of connections, carefully identify the various connections on the figure. Start with pin #1 and work up to pin #8.

This step is important, so take care. Next, we move on to identifying the actual components.

Resistors

Start with the resistors first. There are three resistors used here. The values are

R1 = 100 kohm/color code = brown, black, yellow
R2 = 10 kohm/color code = brown, black, orange
R3 = 1 kohm/color code = brown, black, red

How do you know which end of the color bands to start reading from? There is an additional color band, typically gold, to indicate that the resistors have a 5 percent tolerance. This means that a nominal value of 1 kohm can either be 1 kohm plus 5 percent (that is, 50 ohm), giving a total of 1.050 kohm, or 1 kohm minus 5 percent for a total of 0.950 kohm. The color bands should be read from left to right. Thus for R3 (1 kohm), the bands will be brown, black, red, and finally gold. (Generally speaking, ignore this color when identifying the resistor values.)

In a similar manner, verify the correct values for R1 and R2. To double-check that you have the correct resistors, use your multimeter set to the resistance range, making sure that you have zeroed the meter first (if required) and avoided placing your fingers across the resistor (this will act as a shunt, giving an erroneous reading). Resistors can be placed either way into the circuit board.

Capacitors

Next, move on to identifying the capacitors. C1 has a value of 10 µF and is termed an *electrolytic capacitor*. It is polarity sensitive, which means that it has a positive and negative terminal and must be inserted into the circuit exactly as shown. Look carefully along the body of the capacitor and notice that a distinguishing line of minus signs is marked on one edge of the capacitor. This line of minus signs matches up with one of the leads, indicating that it is the negative terminal. This particular type of capacitor is barrel-shaped and has two leads coming from the same end of the component. The value of the capacitor is actually marked on the body—10 µF.

The next capacitor, C2, is quite different in shape and is called a *ceramic disc capacitor*. The value is printed on the body in a coded form. For the 0.01-µF value, the coded number is 103'. The way this is derived is as follows: 103' is a shortened way of defining the value in picofarads (pF). The last number on the right—3'—tells us how many zeros should be added after the first two numbers. Thus 103' = 10 and three zeros, or 10,000 pF. Picofarads are related to microfarads (µF) by the relation: 1 µF = 1,000,000 pF. Therefore, we can see that 10,000 pF is the same as 10,000 ÷ 1,000,000 µF, or 0.01 µF. This capacitor is not polarity sensitive and can be inserted either way.

Capacitor C3 (0.1 µF) is a disc ceramic type with a coded value of 104'. This code corresponds to 100,000 pF, which is the same as 100,000 pF ÷ 1,000,000 µF, or 0.1 µF.

Capacitor C4 (100 µF) is also an electrolytic type, and it is essential to correctly identify the polarity of the leads. The negative terminal is always connected to ground for the circuit projects described herein.

Integrated Circuit

This is an eight-pin device with the marking 555' on the top. A circle etched onto the package identifies pin #1. The pins are arranged four per side. The numbering scheme for the IC is as follows. With the IC positioned so that the legend reads the correct way up and the circle is positioned at the lower left-hand corner, pin #1 is nearest the circle and, going from left to right, the pin numbers are #2, #3, and #4. Moving upward to the top row of pins, the numbering now runs from right to left—#5, #6, #7, and #8. What you should thus have is as follows: pin #1 at the lower left-hand corner, pin #4 at the lower right-hand corner, pin #5 at the upper right-hand corner, and pin #8 at the upper left-hand corner.

Mechanical Components

There are three mechanical components plus the printed circuit assembly board to complete the list of parts. The 9-volt battery can be purchased anywhere and should be familiar to you as a power source for such everyday items as radios and smoke detectors. There is a positive and negative polarity. Use your voltmeter on the dc voltage range to verify the positive terminal (this is also marked on the battery). The battery snap is fitted onto the battery terminals and has two wires (red for positive and black for negative) leading out from it. The attachment to the battery is made via these wires. Do not connect the battery to the snap while assembling the components. Leave the battery connection until the last step. For testing purposes, however, insert the clip onto the battery and verify that the snap leads do, in fact, correspond with the battery polarities. Take care that the bare ends of the wires do not touch each other; otherwise the battery will be rapidly drained of current. The snap terminals should fit snugly onto the battery; if not, squeeze carefully and slightly with pliers to ensure a tight fit. A loose fit will cause an intermittent connection and lead to no end of problems.

The switch, S1, is a miniature single pole, single throw (SPST) component and has either two or three terminals (depending on the actual type obtained); the difference is not important. For a two-terminal type, use the two terminals; for a three-terminal type, use the center terminal and either one of the other two terminals. The switch can be tested easily. Wire the switch in series with the battery clip and the voltmeter, and you should have the following connections: The black clip wire goes to the black lead of the voltmeter (an alligator clip makes a useful means of temporarily attaching two wires together); the red clip wire goes to one of the switch terminals; and the remaining switch terminal goes to the red voltmeter lead.

Connect the battery. Depending on which way the switch toggle is set, the meter will read either 0 or 9 V. Flip the switch toggle—the meter will change in reading. For your own preference, when you come to mount the switch later, you can choose whether you want the down position to represent on or off.

Assembly Board

NOTE: The layout shown in Figure 17-2 is a suggestion only. There is plenty of space on the assembly board, and for beginners it is best to allow ample space between components. Because of differences in component sizes, feel free to vary the actual layout to suit yourself. Just make sure that the electrical connections are still as shown. Go through this diagram carefully, making sure that you follow each connection point. Match this diagram with the earlier electrical schematic in Figure 17-1.

The custom SINGMIN PCB is a universal hobby project board that greatly improves the chances of having your circuit work the first time it is switched on. Figure 17-2 shows the layout used. Follow this pattern carefully first, before attempting to solder. Place the components in the positions as shown and carefully check that all connections are exactly as indicated. It is vitally important that you have already practiced soldering and can make good solder joints that are shiny, have no excess solder, and do not impart excess heat to the board.

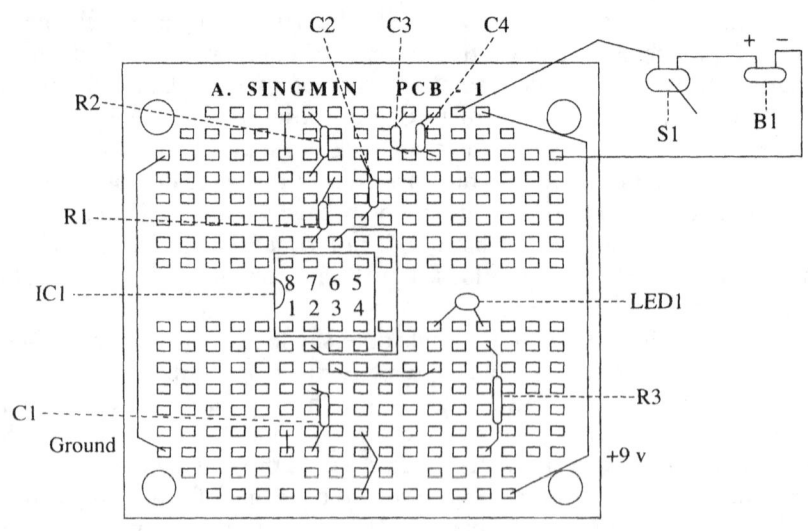

Figure 17-2 Layout for Project #1

Construction Details

Step 1

Start by placing the IC socket in the SINGMIN PCB as shown. The notch in the socket should face left. Bend the leads gently over to lie flush with the board. The socket should now be self-supporting. Start with just two opposite corner leads first. Turn the board over and check that the socket is positioned exactly as shown. Once you are satisfied, bend the rest of the leads in place.

Preheat your soldering iron. Allow time for the tip to heat up and have a moist sponge on hand for periodically cleaning the tip. When ready, tin the tip with solder, wipe it off, and apply the heated, clean tip firmly to the first corner pin #1 (when looking from the top, with the notch in the socket facing left, pin #1 is at the lower left-hand edge). Apply solder to the junction of the tip and terminal. Within a few seconds, the solder will melt.

Remove the solder, then remove the tip and keep the board steady until the solder has cooled (typically 5 to 10 seconds). Turn the board over and check that the socket is straight and is sitting flush with the board. If this is not the case, then the joint can be reheated (without extra solder) and repositioned; however, this step is not critical, and it is better to put up with minor oddities at this stage. There is the danger of imparting excess heat to the components if you are a beginner to soldering.

Finish off the rest of the pins, anchoring the corners first to obtain a good fit to the board. After the socket has been completed, check carefully, if needed with a magnifying glass, that all of the pins have adequate solder, all of the joints are shiny, and there are no solder splashes between adjacent pins to cause a short-circuit. Do not insert the IC at this stage.

Step 2

There are seven solder links to be inserted next. Prepare suitable lengths of solid gauge wire by removing the insulation with wire strippers. The list below shows all of the links to be made. Proceed carefully, soldering one link at a time. Check the integrity of each solder joint before proceeding to the next one. Leave excess wire protruding while making the solder joint. After the solder has cooled, the wire can be trimmed with wire cutters. Do not cut the wire flush with the board; leave a short length showing so as not to impart mechanical stress to the board or components. Refer to Figure 17-2 for the position of the links, which are all marked for clarity.

- Link 1: Upper ground to lower ground
- Link 2: Upper positive supply to lower positive supply
- Link 3: Pin #1 to ground
- Link 4: Pin #4 to positive supply
- Link 5: Pin #8 to positive supply
- Link 6: Pin #2 to pin #6
- Link 7: Pin #3 to LED1

Step 3

Start with LED1 and resistor R3 (1 kohm). Look closely at the LED. There is a flat surface on one edge of the LED showing that the lead nearest this pin should go to the negative supply voltage. Temporarily hook up the resistor R3 to either end of the LED and connect the remaining LED terminal and free resistor end to a 9-volt battery. Note the polarities carefully. If the LED does not light up, then reverse the connections to the battery. Note which end of the LED goes to the positive terminal. This will be the end that will later go to pin #3 of IC1. We are going to locate the LED off the board later, so two short lengths (6 inches or so) of flexible wires are used as extension leads to the LED. Solder the flexible leads to the LED and insulate the bare wires to prevent them from touching and shorting out. Electrical insulation tape can be used.

Do not solder in the extended LED to the SINGMIN PCB at this stage. Resistor R3, however, can be inserted. Bend the resistor leads at right angles to fit the appropriate holes in the board as shown. Do not bend the wires right up to the body of the resistor, but instead use miniature pliers to isolate the bending stresses from the resistor body. Insert the resistor into place and solder. Once cooled, the leads can be trimmed to length.

Step 4

Resistor R1 (100 kohm) is handled in a similar fashion; however, because it goes to two adjacent pins (#6 and #7), there is insufficient space for R1 to be bent in the same way as the first resistor. Instead, bend just one of the leads back on itself so that the two leads face the same direction. Now R1 will easily fit in as shown. Once more, solder the resistor into place, allow it to cool, then check and trim the leads.

Step 5

Resistor R2 (10 kohm) is bent to fit the connection from pin #7 and the positive supply voltage. All of the resistor connections are now completed.

Step 6

Electrolytic capacitor C1 (10 µF) is connected from pin #2 to ground (negative). Observe the polarity carefully. The positive end of the capacitor goes to pin #2, and the negative end goes to ground.

Step 7

Capacitor C2 (0.01 µF) goes from pin #5 to ground and can go in either way because it is not polarity sensitive.

Step 8

Capacitor C3 (0.1 µF) is added across the positive supply line and ground.

Step 9

Capacitor C4 (100µF) is also added across the positive supply line and ground. The negative end of C4 goes to ground.

That completes all of the electronic component connections.

Step 10

The few remaining mechanical parts are connected next. Switch S1 needs to have two flexible wires (6 inches long) soldered to two terminals, as explained earlier. When stripping flexible wire, twist the bare ends together before soldering and tin sparingly with solder. One end of the extended switch connection goes to the positive supply rail as shown. The other end of S1 is soldered to the red (positive) wire on the battery snap. The wires on the battery snap are fragile, and to prevent excess stress on them it's a good idea to extend them as well. As in all cases, insulate any bare soldered wires. The negative (black) end of the battery snap goes to ground. Finally, solder in the LED, making sure that the notched end goes to ground. That's it—the project construction is complete.

Step 11

Perform a thorough check of all of the connections.

Step 12

Take IC1—the 555 timer—and locate pin #1. Position the IC so that the notch (if there is one) faces left, or the identifier circle is in the lower left-hand corner, and the name of the IC reads the correct way up. Place the IC over the socket and you'll see that the leads might be wider than the socket. Gently bend both sets of the IC leads inward until the IC can be inserted into the socket. Make sure all of the pins go in straight and smoothly.

Step 13

Check that the switch, S1, is in the off position. Connect the 9-volt battery. Momentarily flick the switch on. If all is well, the LED will start flashing; however, if there is no sign of life, then switch off immediately, disconnect the battery, and start rechecking the circuit connections. Wrong connections, poor solder joints, and short-circuits are the most common causes of problems. Component failure is rarely, if ever, to blame.

Conclusion

If after construction the project works well, then you can purchase a plastic hobby project case from Radio Shack and mount all of the items inside. I also advise that you keep this first project as a valuable reference so that you can refer to it for future projects. Choose a larger rather than smaller case because this will make the working process much easier. The following projects will have a less-detailed description of the basics we have just covered; it is assumed that you have completed Project #1.

Project #2: Variable Low-Frequency LED Flasher/Driver

This second project builds on the experience gained from Project #1, and more components are added to expand on the functionality. There is now a control for varying the frequency, a switch to select a high or low range (for frequency), and an output jack socket that enables you to use the output as a trigger source. For example, if you wanted to trigger a high-current filament bulb to act as a strobe, then a relay and a buffer amplifier would be needed to drive the filament bulb. The trigger for the buffer amplifier would be provided by the output of this variable-frequency flasher/driver.

Although the initial construction phase of this project is similar to the previous one, the construction details are repeated here in full. This project will also introduce you to working with several new mechanical components.

Circuit Description

This project's adaptation of the circuit in Project #1 provides additional capability to vary the pulse rate of the 555 timer oscillator and provides a useful takeoff point for the pulse signals via a jack socket. Frequency variation is provided by a potentiometer that varies the timing resistance, and a selector switch controls the choice of timing capacitor.

Parts List

Semiconductor
IC1: LM 555 timer

Resistors
R1 = 10 kohm
R2 = 100 kohm potentiometer
R3 = 10 kohm
R4 = 1 kohm

Capacitors
C1 = 10 µF
C2 = 100 µF
C3 = 0.01 µF
C4 = 0.1 µF
C5 = 100 µF

Additional Parts and Materials

LED1: Light-emitting diode
S1: Miniature SPDT toggle switch
S2: Miniature SPST toggle switch
J1: Miniature 1/8-inch jack socket
J2: Miniature 1/8-inch jack plug (needed for testing purposes only)
B1: 9-volt battery
9-volt battery snap
8-pin IC socket
Knob to fit potentiometer
General purpose circuit assembly board
Hook-up wire (solid and stranded)

Pin Connections (see Figure 17-3)

Pin #1

This pin is coupled directly to ground with a jumper wire.

Pin #2

This pin is cross-connected to pin #6 and to the center terminal of two-way selector switch S1, which will select either capacitor C1 (10 µF) or C2 (100 µF), thus providing a selection of frequency ranges (high or low).

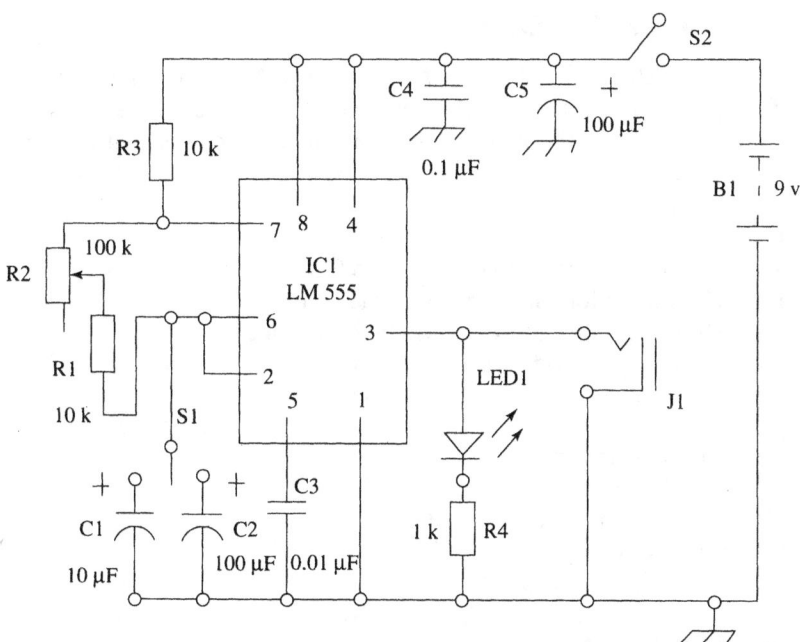

Figure 17-3 Project #2: Variable low-frequency LED flasher/driver

Pin #3
The output signal from this pin drives a light-emitting diode, LED1, and is terminated in a miniature jack socket, J1, to provide a connection to the outside world. Resistor R4 (1 kohm) is a current-limiting resistor for LED1.

Pin #4
This pin is connected to the positive supply voltage.

Pin #5
This pin is grounded through a capacitor, C3 (0.01 µF).

Pin #6
This pin is connected to pin #2. Pin #6 also is coupled to pin #7 via a combination of series resistor R1 (10 kohm) and potentiometer R2 (100 kohm). R1 is a fixed resistor, whereas R2 is a variable potentiometer that allows for a continuous adjustment in frequency.

Pin #7
This pin is taken to the positive supply rail via resistor R3 (10 kohm).

Pin #8
This pin is always taken to the positive supply rail.

Component Identification

Figure 17-3 shows the electrical schematic for the variable low-frequency LED flasher/driver. Identify on the schematic all of the connections given above. Next, match up the components you have purchased with the parts listed in the schematic.

Resistors
The resistor values should match up as follows:
R1 = 10 kohm/color code = brown, black, orange
R3 = 10 kohm/color code = brown, black, orange
R4 = 1 kohm/color code = brown, black, red

Capacitors
C1 = 10 µF/electrolytic
C2 = 100 µF/electrolytic
C3 = 0.01 µF/number code = 103'
C4 = 0.1 µF/number code = 104'
C5 = 100 µF/electrolytic

Integrated Circuit
This is an eight-pin device with the marking 555' on the top of the device. A circle near the lower left-hand corner identifies pin #1.

Mechanical Components

The 9-volt battery, battery snap, and on/off power switch S2 are as used previously. The new components begin with the potentiometer R2 (100 kohm). A potentiometer has three terminals. By rotating the shaft, the resistance between the center terminal and either of the outer two will vary. Set your multimeter to the resistance range and temporarily connect the center terminal and one of the outer terminals to the meter. By rotating the shaft, you can see the resistance change from 0 ohm to 100 kohm. Keep the center connection intact, but connect the other outer terminal. Notice that the direction of resistance change is now reversed as the shaft is rotated. For the application here, we want only the resistance to be varied, so only two terminals need be used.

In this circuit, we want the frequency to increase as the control is rotated clockwise, which means that the resistance must decrease as the shaft is rotated. Identify, in turn, which of the arrangements will give you this result. When viewed from the front of the potentiometer, is it the center and left-hand terminal or the center and right-hand terminal? It should be the center and left-hand terminal. Mark that terminal when you've found it. We will be using that one later (along with the center terminal).

The two-way selector switch S1 has three terminals and is used to switch the center pin to either of the two outer pins, depending on which way the switch is toggled. Check this out first. Connect one end of the meter, which should be set to the resistance range, to the center pin; connect the remaining meter lead to either of the outer pins. Take note in which direction the switch has to be toggled in order for a connection to be made. Keep the center connection intact and now reverse the outer connection. The toggle direction needed to make this new connection will be the opposite way. What you've just verified is that the switch will alternatively connect the center pin to either of the two outer pins; in other words, this is a selector switch.

Assembly Board

NOTE: The layout shown in Figure 17-4 is a suggestion only. There is plenty of space on the assembly board, and for beginners it is best to allow ample space between components. Because of differences in component sizes, feel free to vary the actual layout to suit yourself. Just make sure that the electrical connections are still as shown. Go through this diagram carefully, making sure that you follow each connection point. Match this diagram with the earlier electrical schematic in Figure 17-3.

Construction Details

Step 1

The first task will be to prepare potentiometer R2. Potentiometers as procured are supplied with a shaft that needs to be reduced in length to fit the

56 BEGINNING ANALOG ELECTRONICS THROUGH PROJECTS

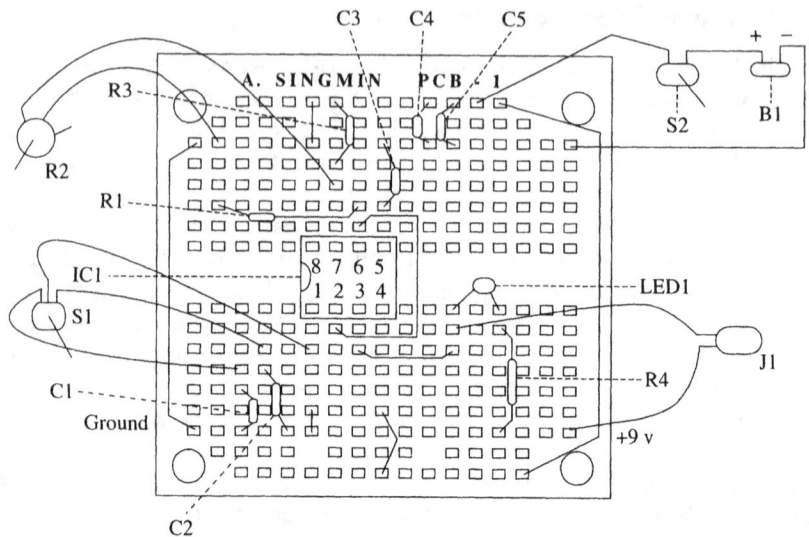

Figure 17-4 Layout for Project #2

type of control knob you will be using. Buy the knob first before cutting down the shaft because the length to be eliminated will depend on the size of knob used. Looking at the potentiometer with the shaft facing you, there will be a small metal tab on the right-hand side. This is an additional securing tab that is not needed for our purposes, and you can use a pair of pliers to break it off. Place the control knob over the end of the shaft and estimate how much of the shaft needs to be cut off. Place the shaft firmly in a vice (never clamp the body because there could be excessive stress) and carefully cut off the excess shaft with a small hacksaw. Smooth off the sharp ends if necessary. Place the knob over the shaft and lock it into place. You will need a small-tipped screwdriver to adjust the securing screw for the knob. Solder two flexible lengths (6 inches) of wire to the two terminals, as described in Project #1.

The jack socket J1 needs to be prepared next. You will also need jack plug J2 and a multimeter set to the resistance range. Jack socket J1 has three terminals. There is one closest to the front of the socket where the plug is inserted and two farther at the back. With the socket positioned so that the terminal closest to the entry point faces upward, the other terminal to be used is the one closest to the bottom of the socket. To verify these terminals, take the jack plug and unscrew the barrel. You will see two terminals inside. A shorter section is connected to the tip of the plug (this should be marked by you as the input terminal), and a longer section is connected to the main body of the plug (which should be marked by you as the ground terminal).

Use your ohmmeter to verify these connections first. Next, insert jack plug J2 (with the barrel still removed) into jack socket J1. Verify with the ohmmeter that the tip of jack plug J2 is connected to the rear jack socket J1 terminal. Mark this terminal on the socket as the input terminal. Next verify that the remaining jack socket J1 terminal is the ground connection. Mark the terminal as such. These identifying markings are critically important in preventing circuit malfunctions later on. Solder two flexible lengths of wire to these two terminals. If you have an assortment of wire colors from which to choose, it is better to use red for the input and black or green for the ground connections. This system saves confusion later. Otherwise, place labels on the wires if they are the same color.

Switch S1 needs three flexible leads (each 6 inches in length) soldered on first. Use your ohmmeter to identify and label which connection corresponds to each switch position. Switch S2 has two flexible leads (again, each 6 inches in length), one of which is soldered to the center terminal and one to either of the outer terminals. Mark the on direction on S2, so you know when power is applied to the circuit.

The LED is identified for the positive and negative polarities and flexible leads (6 inches in length) are soldered on.

All flying flexible wire ends must be twisted together before soldering.

Step 2
The IC socket is located in the SINGMIN PCB as per Figure 17-4. Bend the leads of the socket inward in order to locate the device more easily into the board. Solder into place and verify the integrity of all solder joints as you go along. Conduct a detailed inspection with a magnifier if necessary.

Step 3
Start with the solid wire links. These links run to the following points:
Link 1: Upper ground to lower ground
Link 2: Upper positive supply to lower positive supply
Link 3: Pin #1 to ground
Link 4: Pin #2 to pin #6
Link 5: Pin #4 to positive supply
Link 6: Pin #8 to positive supply
Link 7: Pin #3 to LED
Link 8: Pin #3 to jack socket J1

Step 4
Start with the LED section. The LED will actually be added at the final stages of assembly, but the resistor R4 (1 kohm) goes in now. Bend the resistor leads to conform to the position shown on the layout diagram. One end of R4 is tied to the ground rail. Ultimately, LED1 will make contact with R4 and pin #3. Trim the leads to length once the solder has cooled.

Step 5
Resistor R1 (10 kohm) and potentiometer R2 (100 kohm) are located in series between pins #6 and #7. Insert R1 first. Note that one end of R1 goes to

pin #6, whereas the other end terminates at an isolated solder site. From this point, one of the flying wires from R2 will later be terminated. The remaining wire from R2 will eventually go to pin #7. All mechanical components are attached at the final stages to make initial assembly easier.

Step 6
Resistor R3 (10 kohm) goes from pin #7 to the positive supply.

Step 7
Capacitors C1 (10 µF) and C2 (100 µF) form part of the selector network. They are physically located in a different position on the SINGMIN PCB, near the lower extremity of the board. Note that both of the negative ends of C1 and C2 terminate at the ground line. The opposite ends of C1 and C2, though, are totally isolated. From these two isolated points, the selector switch S1 will eventually make a connection via the outer two switch terminals. Pin #2 eventually will go to the center switch S1 terminal. Be aware of the polarity of C1 and C2 (the negative ends go to ground) when installing the electrolytic capacitors.

Step 8
Capacitor C3 (0.01 µF) runs from pin #5 to ground. This capacitor, a disc ceramic type, can be inserted either way.

Step 9
Capacitor C4 (0.1 µF) is added across the supply and ground lines.

Step 10
Capacitor C5 (100 µF) is also added across the supply and ground lines.

Step 11
At this stage, without the encumbrances of switches and potentiometer, it is easier to perform a thorough circuit check. Be absolutely thorough. Checking becomes more difficult as the assembly proceeds. If everything checks out, then let's proceed with the construction.

Step 12
Insert power switch S2 as shown (one end to the positive supply rail). This switch is connected in series to the positive (red) wire from the battery snap. The flying black (negative) battery clip wire is taken to the ground line.

LED1 goes into the board between pin #3 and resistor R4 (1 kohm). Watch the polarity for LED1 (the negative flat goes to resistor R4).

Selector switch S1 has the center terminal routed to pin #2 and the remaining terminals to C1 (10 µF) and C2 (100 µF). It doesn't matter which way they go.

Jack socket J1 goes in next to pin #3 and ground. Make sure that the ground terminal of J1 goes to the common ground rail.

Step 13

Carefully insert the IC into the socket, bending the leads carefully if needed. Make sure the polarity is correct, with pin #1 on the IC matching pin #1 of the socket.

Step 14

Check that the power switch S2 is in the off position. Rotate the potentiometer to the extreme counterclockwise position. Attach the 9-volt battery. Switch it on momentarily and watch the LED for signs of life. All being well, the LED will flicker. Advance the potentiometer clockwise. The frequency will increase. Toggle S1. The frequency will either decrease or increase depending on which capacitor has been selected. In the event of a dead circuit, switch it off, disconnect the battery, and start checking again for errors in assembly. Errors are usually the result of mistakes in component placement and/or poor soldering (short-circuits, open circuits).

Assuming all is well, the output from the jack socket can also be verified. Take the previously used jack plug J2 and solder two flexible wires to the terminals. Use red for the center input pin and black for the outer ground terminal. Connect the red wire to your multimeter set to the dc volts range (able to display 9 volts) and the black wire to the negative terminal. Reconnect the battery. Plug in the jack plug. Rotate the control fully counterclockwise and select the slower frequency range. Switch it on. The meter needle will oscillate in step with the LED, showing that you now have an external feed.

Conclusion

Purchasing a plastic project case makes an excellent conclusion to this project. Drill holes to match a nice, even layout and add labels to suit in order to finish up with an attractive and versatile working unit.

Project #3: Fixed Low-Gain Audio Power Amplifier

One of the projects that will almost always be used is a simple audio power amplifier capable of driving a speaker. Speakers are characterized by having a very low impedance or resistance, typically around 8 ohm. In order to drive a speaker from an audio source, you need to isolate the low speaker impedance by using an audio isolator such as a low-gain power amplifier. Fortunately, there is a superb, easy-to-use integrated circuit (IC), the LM 386, that requires only a mere handful of components to get going. Power comes from a regular alkaline 9-volt battery. Although the rated power is less than 1 watt, there is more than enough drive capability for most applications.

Circuit Description

The LM 386 is basically a four-terminal device. There are the customary power (terminal A) and ground (terminal B) connections, plus the input (terminal C) and output (terminal D). Because the intended use is for audio applications, this amplifier is called an *alternating current* (ac) amplifier, which means that capacitors (which block dc and permit ac) must be present at the input and output. The signal gain is low (×20), and hence a reasonably high signal—for example, that from a portable radio or cassette recorder—must be used.

We really want this amplifier for the output capability, however, rather than for providing signal gain. A speaker or headphones has a low impedance, typically around 8 ohm. In order to drive such a device, a power driver or amplifier must be used. Placing a speaker directly across a signal source causes the signal to be significantly attenuated because of the loading effect. Power is from a 9-volt battery.

Parts List

Semiconductor
IC1: LM 386 audio power amplifier

Resistors
R1 = 2.7 kohm

Capacitors
C1 = 0.1 µF
C2 = 0.1 µF
C3 = 100 µF
C4 = 100 µF

Additional Parts and Materials
LED1: Light-emitting diode
S1: Miniature SPST toggle switch
J1: Miniature 1/8-inch jack socket
SPKR: Miniature 8-ohm 2-inch speaker
B1: 9-volt battery
8-pin IC socket
9-volt battery snap
General purpose circuit assembly board
Hook-up wire (solid and stranded)

Pin Connections (see Figure 17-5)

Pin #1
There is no connection to this pin.

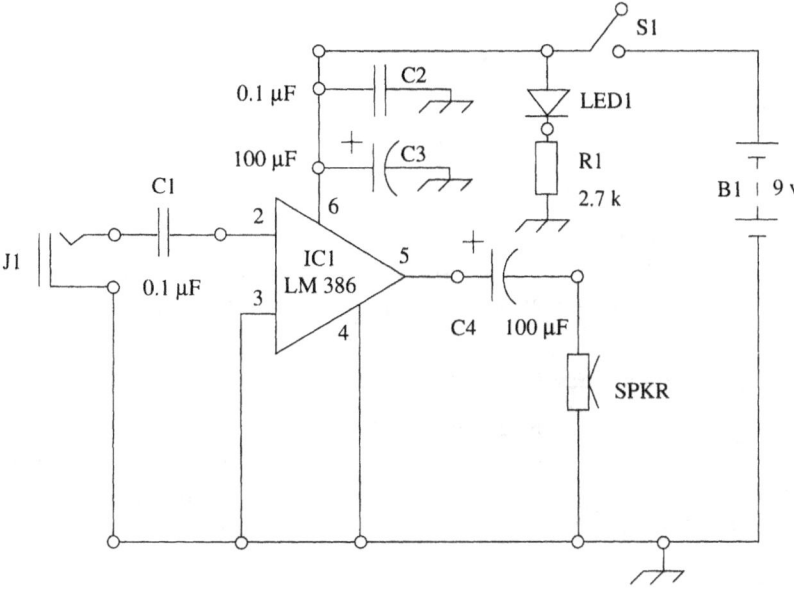

Figure 17-5 Project #3: Fixed low-gain audio power amplifier

Pin #2

This is the signal input pin and, because this is an audio signal amplifier, the input signal is always ac coupled. That means there will always be a capacitor in series between the input signal and pin #2, hence the presence of capacitor C1. A value of 0.1 µF is a good standard starting value for most applications.

Pin #3

This pin is taken to ground.

Pin #4

This pin is taken to the official ground connection.

Pin #5

The output signal is taken from this pin. Again, for audio applications, the signal is always ac coupled, hence the presence of capacitor C4, which typically is 100 µF and is thus an electrolytic type. Note the polarity, with the positive end going to pin #5.

Pin #6

This pin goes to the positive supply voltage. Disc ceramic capacitor C2 (0.1 µF) and electrolytic capacitor C3 (100 µF) provide a circuit-smoothing function. They are wired across the positive supply and ground and are positioned close to the IC. C3's positive end matches with the positive supply rail.

Pin #7
There is no connection to this pin.

Pin #8
There is no connection to this pin.

LED1 and resistor R1 (2.7 kohm) are wired across the positive supply line and ground and indicate when the power is on.

Component Identification

Figure 17-5 shows the complete electrical schematic for the fixed low-gain audio power amplifier. Using the list above, identify carefully all of the connections. After you have done this, the next stage involves identifying and matching up the actual components with the schematic.

Resistors
Only one resistor is used for this project. Resistor R1 has a value of 2.7 kohm and can be identified as follows:
R1 = 2.7 kohm/color code = red, violet, red

Capacitors
C1 = 0.1 µF/number code = 104'
C2 = 0.1 µF/number code = 104'
C3 = 100 µF/electrolytic
C4 = 100 µF/electrolytic

Integrated Circuit
The LM 386 is in an eight-pin package. The pin numbering scheme is standard, with the circle locating pin #1.

Mechanical Components

The stock battery snap, 9-volt battery, and power switch S1 are the same as we've been using before. The signal input is fed via a miniature mono jack socket, J1. The speaker, SPKR, is a small two-inch type (the actual size is not critical—use whatever you can find). For convenience, the speaker will be mounted on flying flexible leads (for later mounting inside a project case). Because the solder terminals located on the back of the speaker are fragile, make the solder quickly and then anchor the extension wires down so that you don't accidentally pull the wires off. The light-emitting diode (LED) makes a convenient indicator light for when the power is switched on. In that way, you won't accidentally leave the amplifier on and drain the battery.

Assembly Board

NOTE: The layout shown in Figure 17-6 is a suggestion only. There is plenty of space on the assembly board, and for beginners it is best to allow ample space between components. Because of differences in component sizes, feel free to vary the actual layout to suit yourself. Just make sure that the electrical connections are still as shown. Go through this diagram carefully, making sure that you follow each connection point. Match this diagram with the earlier electrical schematic in Figure 17-5.

Construction Details

Step 1

Insert the IC socket into the SINGMIN PCB as shown. Bend the socket leads back against the board to anchor the device. Check for correctness before proceeding to solder. Verify the integrity of each pin after soldering before proceeding to the next.

Step 2

A series of shorting wire links provides the main interconnection structure for the amplifier. Use the list below and check each carefully as you proceed.

Link 1: Upper ground to lower ground
Link 2: Pin #3 to ground
Link 3: Pin #4 to ground
Link 4: Pin #6 to positive supply

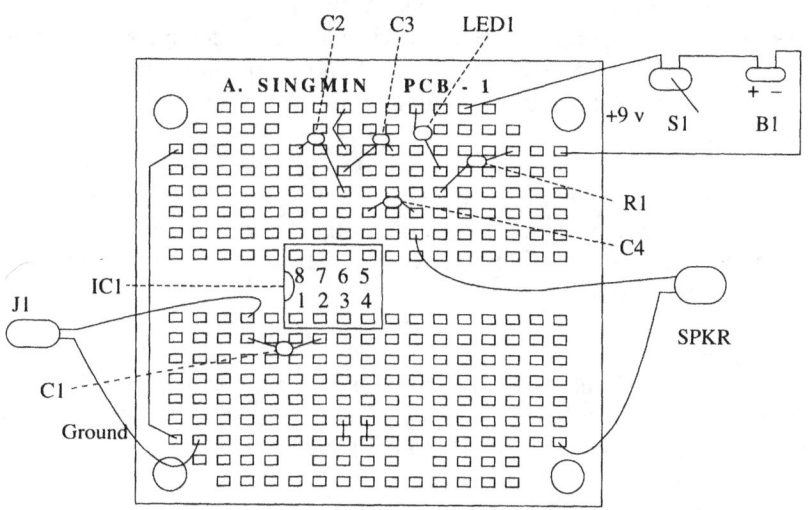

Figure 17-6 Layout for Project #3

Step 3
Insert disc ceramic capacitor C1 (0.1 µF) into the board as shown. C1 goes between pin #2 and a floating solder point on the board. This takeoff point will eventually terminate in a mono jack socket, J1.

Step 4
Capacitor C4 (100 µF) feeds the output from pin #5 to the speaker. Check the polarity matches, ensuring that the positive capacitor end goes to pin #5. The negative end is terminated for the time being in an isolated solder point. Ultimately, the speaker will be joined here.

Step 5
Capacitor C2 (0.1 µF) disc runs between pin #6 and the positive rail. From the same pin, there is also an electrolytic capacitor, C3 (100 µF). Watch the polarity—positive to positive.

Step 6
Finally, add the LED and resistor R1 (2.7 kohm).
The power-on indicator, LED1 with its current-limiting resistor, is tied across the positive supply and ground lines. Match the LED polarity correctly, with the flat going to ground.

Step 7
Conduct a thorough check for component placement accuracy and solder integrity. All being well, proceed onward.

Step 8
Add the power switch S1 first. One terminal of the switch goes to the positive supply voltage point on the board via a length of flexible wire. The other switch end is soldered to the red (positive) end of the battery snap. The black (negative) battery snap lead is terminated at the ground rail via a length of flexible wire.

Step 9
Add the input jack socket J1 to the isolated end of capacitor C1 (0.1 µF) and ground. Make sure that the jack socket polarity is correctly matched to the input and ground connections.

Step 10
The speaker should be handled carefully to prevent any flying leads from being pulled off. It can be connected either way into the circuit. The connections are to the negative end of C4 (100 µF) and ground.

Step 11
Insert the IC gently into the socket, bending the leads inward if needed, matching pin #1 of the IC and socket.

Step 12

Check that the switch S1 is in the off position. Add the 9-volt battery to the battery snap. Momentarily switch it on. The power LED should light. Put your ear to the speaker face. You might be able to detect a slight hiss. Place your finger on the input terminal of J1. The hiss should increase in level. This is all normal. Should there be no signs of life, switch it off, remove the battery, and recheck all connections.

Step 13

If all is well at this stage, then the final step is to prepare the signal feed to the input. You can use either an off-the-shelf cable assembly with a mono miniature 1/8-inch jack plug at either end or make one up yourself. If you're buying one, be absolutely sure that the plugs are mono 1/8-inch jack plugs and not stereo. Stereo plugs will not work here.

An ordinary, inexpensive am radio will serve as a useful test source. The cost is very little even if you have to buy one. Tune in to a station and set the volume very low so that it's just audible. Switch off the radio. Make sure the amplifier is switched off.

Connect the cable from the earpiece output socket of the radio to the input of the amplifier. First switch on the amplifier. Next switch on the radio. All being well, the sound, very much louder because of the ×20 gain from the amplifier, should now be coming from your amplifier speaker. Adjust the radio's volume control to suit. The system is now complete. Always switch off in the reverse order; that is, radio first, then amplifier.

Conclusion

A project case nicely completes this project. The speaker can be attached with a few squeezes of a glue gun. Drill out a few holes first to allow the sound to exit properly. Labels will add to the finish.

Project #4: Fixed-Frequency Audio Tone Generator

A simple project for testing audio equipment is the tone generator, which is capable of sending a tone through, say, an amplifier for quick evaluation. The frequency is within the audio frequency range so as to be audible and is of a low amplitude in order not to overload the input stages.

Circuit Description

This circuit uses the 555 timer configured as a free-running, fixed-frequency square wave oscillator. The selection of the output frequency is positioned in the audio frequency band by two timing resistors and a timing capacitor. Power is from a 9-volt battery.

Parts List

Semiconductor
IC: LM 555 timer

Resistors
R1 = 10 kohm
R2 = 1 kohm
R3 = 100 kohm
R4 = 10 kohm
R5 = 2. kohm

Capacitors
C1 = 0.1 µF
C2 = 0.01 µF
C3 = 0.1 µF
C4 = 0.1 µF
C5 = 100 µF

Additional Parts and Materials
LED1: Light-emitting diode
S1: Miniature SPST toggle switch
B1: 9-volt battery
J1: Miniature 1/8-inch jack socket
9-volt battery snap
8-pin IC socket
General purpose circuit assembly board
Hook-up wire (solid and stranded)

Pin Connections (see Figure 17-7)

Pin #1
This pin goes to ground using a jumper wire.

Pin #2
A jumper cross-links this pin #2 to pin #6. A capacitor, C1 (0.1 µF), one of the timing components, runs from pin #2 to ground. A resistor, R1 (10 kohm), couples pin #6 to pin #7.

Pin #3
The signal tone output exits from this pin. Because the signal amplitude is large at this point, two resistors, R3 (100 kohm) and R4 (10 kohm), act as a signal attenuator and reduce the voltage by a factor of 10, to a more useful level. The reduced signal is fed through a capacitor, C3 (0.1 µF), and terminates in a jack socket, J1.

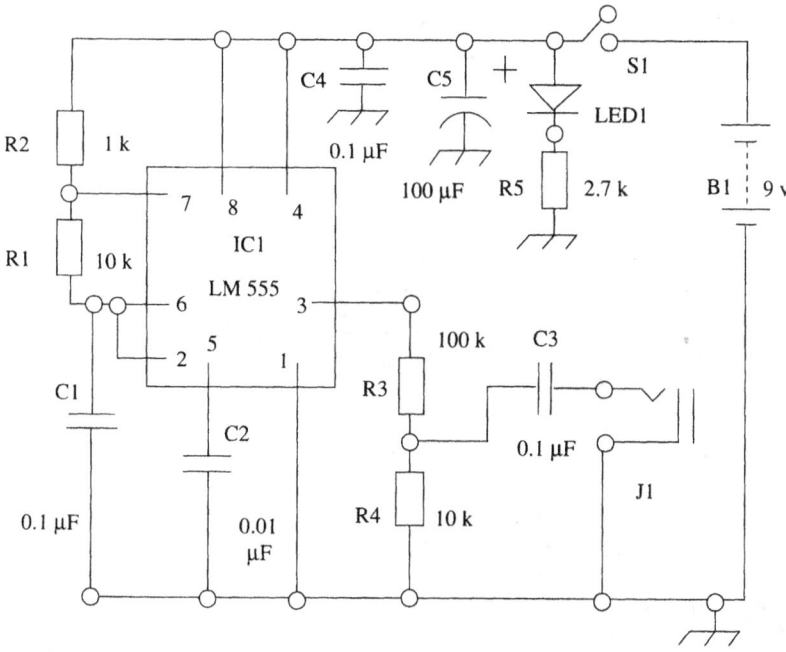

Figure 17-7 Project #4: Fixed-frequency audio tone generator

Pin #4
This pin is connected to the positive supply voltage line.

Pin #5
This pin is taken to ground via capacitor C2 (0.01 µF).

Pin #6
This pin is connected to pin #2. A resistor, R1 (10 kohm), runs to pin #7.

Pin #7
This pin is taken to the positive supply rail via resistor R2 (1 kohm).

Pin #8
This pin is linked to the positive supply voltage. A combination light-emitting diode (LED1) and resistor R5 (2.7 kohm) is wired across the supply voltage to act as a power-on indicator.

Component Identification

Figure 17-7 shows the complete electrical schematic for the fixed-frequency audio tone generator. It is a good idea to start with identifying, on the schematic, all of the pin connections listed above. Once you have done this, go to the actual components you have purchased and match up each of the items.

Resistors

Five resistors are used in this project. Look at the color codes below and carefully match them up.

R1 = 10 kohm/color code = brown, black, orange
R2 = 1 kohm/color code = brown, black, red
R3 = 100 kohm/color code = brown, black, yellow
R4 = 10 kohm/color code = brown, black, orange
R5 = 2.7 kohm/color code = red, violet, red

Capacitors
C1 = 0.1 µF/number code = 104'
C2 = 0.01 µF/number code = 103'
C3 = 0.1 µF/number code = 104'
C4 = 0.1 µF/number code = 104'
C5 = 100 µF/electrolytic

Integrated Circuit

The familiar 555 timer integrated circuit (IC) is an eight-pin device with the marking 555' on the top surface. Pin #1 is identified with the imprinted circular mark near the lower left-hand corner.

Mechanical Components

The switch S1, battery snap, jack socket J1, and 9-volt battery make up the mechanical components list.

Assembly Board

NOTE: The layout shown in Figure 17-8 is a suggestion only. There is plenty of space on the assembly board, and for beginners it is best to allow ample space between components. Because of differences in component sizes, feel free to vary the actual layout to suit yourself. Just make sure that the electrical connections are still as shown. Go through this diagram carefully, making sure that you follow each connection point. Match this diagram with the earlier electrical schematic in Figure 17-7.

Construction Details

Step 1

Locate the IC socket in the SINGMIN PCB using Figure 17-8 as a guide. The socket notch must face toward the left-hand side of the board.

Construction Details for 10 Simple Projects 69

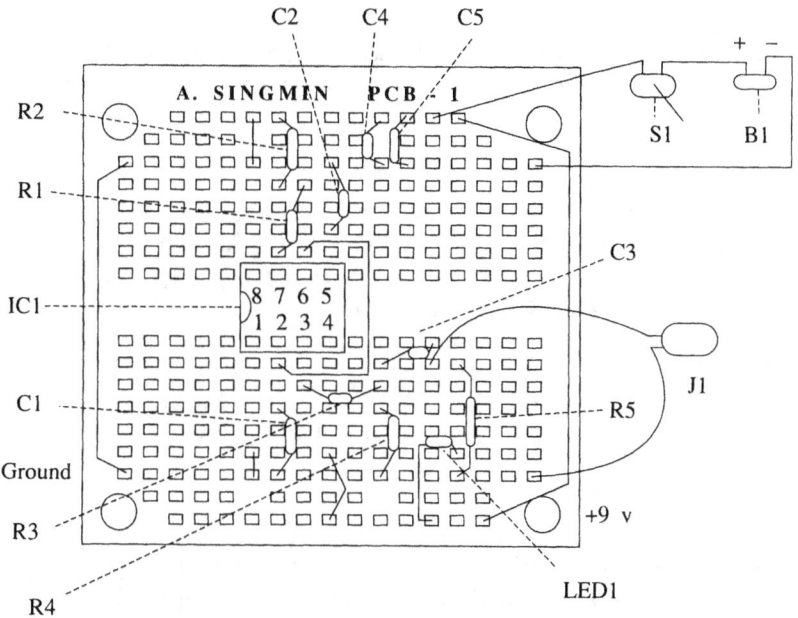

Figure 17-8 Layout for Project #4

Verify that all solder joints are error free and the socket seats evenly to the board before proceeding. Inspect all connections carefully with a magnifying aid.

Step 2
Several wire links define the initial interconnection scheme. Solid wire links are used here. The following links are needed:

Link 1: Upper ground to lower ground
Link 2: Upper positive supply to lower positive supply
Link 3: Pin #2 to pin #6
Link 4: Pin #4 to positive supply
Link 5: Pin #8 to positive supply
Link 6: Jack socket to output signal

Step 3
Start with the capacitors. C1 (0.1 µF) is soldered between pin #2 and ground. With the position shown, there is a nice, easy fit. Trim component leads to length after first checking for correctness. Capacitor C2 (0.01 µF) is added in the same way, running this time from pin #5 to ground. The use of the ground link can now be seen, making it considerably easier to locate components on both sides of the IC.

Step 4
Locate resistor R1 (10 kohm) between pin #6 and pin #7. The fit is tight, so take care in soldering the two adjacent tracks. Resistor R2 (1 kohm) is connected between pin #7 and the positive supply rail.

Step 5
Pin #3 is connected to a resistor, R3 (100 kohm), that itself is terminated at a floating point. The second resistor, R4 (10 kohm), couples the free end of R3 to ground. At the junction of R3 and R4, a final capacitor, C3 (0.1 µF), provides the ac coupling needed for the output.

Step 6
The light-emitting diode, LED1, and associated resistor R5 (2.7 kohm) are located across the positive supply rail and ground. When power is applied, the LED will light up.

Step 7
Add capacitors C4 (0.1 µF) and C5 (100 µF) across the positive supply and ground. The positive end of C5 goes to the positive supply line.

Step 8
That completes the assembly of all the electronic components. Thoroughly check all connections now, especially solder joint validity.

Step 9
Switch S1 has one lead connected to the positive supply rail via a short length of flexible wire. The other switch end is coupled to the red (positive) battery snap lead. The black (negative) battery snap lead goes to ground.

Step 10
Attach jack socket J1 to the free end of capacitor C3 (0.1 µF), making sure that the polarity for the socket is correct.

Step 11
With the switch S1 in the off position, insert a 9-volt battery into the battery snap and momentarily switch it on. The LED should light. That being well, switch it off and couple the output signal into a suitable test amplifier. The previous project (fixed low-gain audio power amplifier) is an ideal test vehicle. First, complete all of the couplings before switching either unit on. Switch the amplifier on first, then switch on the tone generator. You should hear a loud, clear audio tone coming from the speaker. If there is no sound, switch off the tone generator first, then the amplifier, and go back to checking all connections.

Conclusion

This test tone generator is an ideal piece of equipment for checking out audio equipment, from speakers to amplifiers. Mount all of the components in a project case to suit your personal preferences.

Project #5: Variable-Gain Audio Power Amplifier

Having completed the basic amplifier in Project #3, in this project we boost the signal gain from ×20 to ×200 and, because this very high gain will be too much in some cases, a volume control is also added to the input. The application is still for audio purposes, hence the input and output signals are ac coupled via capacitors. The speaker still remains as an on-board miniature type. Choose the physical speaker size to match the final project case you want to use. A two-inch diameter speaker is a good starting size.

Circuit Description

The LM 386 is once again used as the popular workhorse for constructing a 9-volt, battery-driven audio power amplifier capable of driving low-impedance speaker loads. The voltage gain is set to ×200, and the addition of an output-compensating network (called a *Zobel network*) improves the high-power performance. An input volume control provides user control over the gain. Radio frequency (RF) breakthrough, which is unwanted pickup of radio station interference, is eliminated with a small RF-bypass capacitor. Power is from a 9-volt battery.

Parts List

Semiconductor
IC1: LM 386 audio power amplifier

Resistors
R1 = 10 kohm potentiometer
R2 = 10 ohm
R3 = 2.7 kohm

Capacitors
C1 = 0.1 µF
C2 = 1,000 pF
C3 = 10 µF
C4 = 0.1 µF
C5 = 100 µF
C6 = 0.1 µF
C7 = 100 µF

Additional Parts and Materials
LED1: Light-emitting diode
S1: Miniature SPST toggle switch
J1: Miniature 1/8-inch jack socket
SPKR: Miniature 8-ohm 2-inch speaker
B1: 9-volt battery
9-volt battery snap
8-pin IC socket
General purpose circuit assembly board
Hook-up wire (solid and stranded)

Pin Connections (see Figure 17-9)

Pin #1
Capacitor C3 (10 µF) links pin #1 to pin #8 to provide the boost in gain from ×20 to ×200.

Pin #2
This signal input pin is fed from the center terminal of potentiometer R1 (10 kohm). A small capacitor, C2 (1,000 pF) shunts pin #2 to ground, preventing any RF breakthrough (unwanted radio station transmissions) that might occur.

Pin #3
This pin goes to ground.

Pin #4
This pin goes to ground.

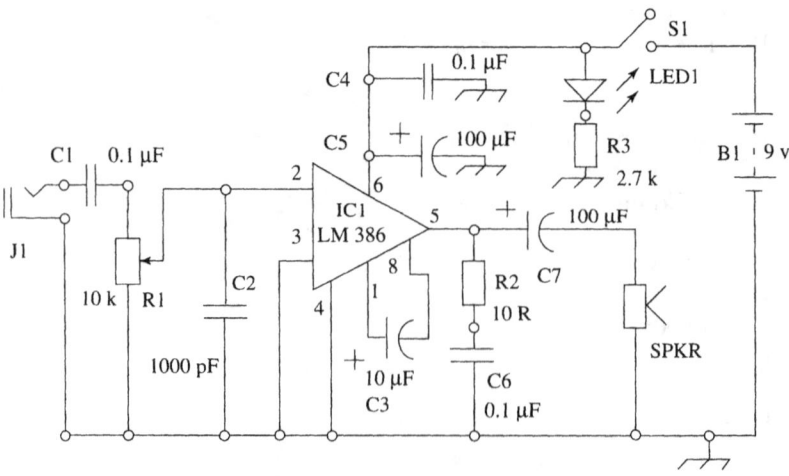

Figure 17-9 Project #5: Variable-gain audio power amplifier

Pin #5
The output signal exits from this pin. A series combination of resistor R2 (10 ohm) and capacitor C6 (0.1 µF) form a Zobel network, which eliminates distortion in the event of very large signals passing through the amplifier. Capacitor C7 (100 µF) leading to the speaker, SPKR, also exits from this pin.

Pin #6
This pin goes to the positive supply rail. Capacitors C4 (0.1 µF) and C5 (100 µF) add stability to the operation of the integrated circuit (IC).

Pin #7
There is no connection to this pin.

Pin #8
This pin is connected to pin #1 via capacitor C3 (10 µF) for the high-gain setting function.

The input terminal to potentiometer R1 (10 kohm) is fed from capacitor C1 (0.1 µF).

The light-emitting diode LED1 and associated resistor R3 (2.7 kohm) act as the power-on indicator and go between the positive supply voltage and ground.

Component Identification

Figure 17-9 shows the electrical schematic for the variable-gain audio power amplifier. Identify all of the connections given above. Match all of the actual components.

Resistors
The resistor values should be matched up as follows:
R2 = 10 ohm/color code = brown, black, black
R3 = 2.7 kohm/color code = red, violet, red

Capacitors
C1 = 0.1 µF/number code = 104'
C2 = 1,000 pF/number code = 102'
C3 = 10 µF/electrolytic
C4 = 0.1 µF/number code = 104'
C5 = 100 µF/electrolytic
C6 = 0.1 µF/number code = 104'
C7 = 100 µF/electrolytic

Integrated Circuit
This is an eight-pin device with the marking LM 386' on the top. A circle near the lower left-hand corner identifies pin #1.

Mechanical Components

Switch S1, speaker, SPKR, battery snap, and 9-volt battery make up the more obvious mechanical components and do not require any further explanation. A potentiometer, R1 (10 kohm), provides the volume control function. The input signal is fed via a miniature jack socket, J1.

Assembly Board

NOTE: The layout shown in Figure 17-10 is a suggestion only. There is plenty of space on the assembly board, and for beginners it is best to allow ample space between components. Because of differences in component sizes, feel free to vary the actual layout to suit yourself. Just make sure that the electrical connections are still as shown. Go through this diagram carefully, making sure that you follow each connection point. Match this diagram with the earlier electrical schematic in Figure 17-9.

Construction Details

Step 1

Prepare the potentiometer, R1 (10 kohm), by first sizing the shaft to fit the control knob. Cut the shaft to length with a small hacksaw, securing the shaft in a vice for stability. Smooth off any rough edges. Make sure that no metal filings fall into the body of the potentiometer. Fit on the control knob.

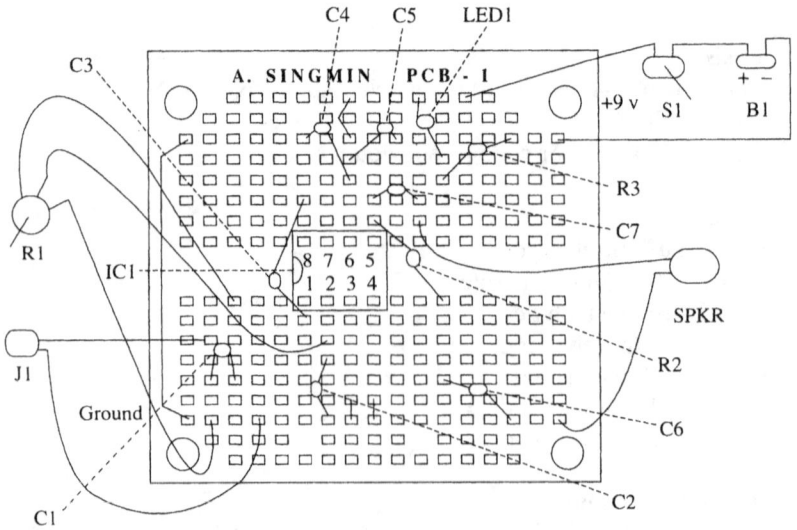

Figure 17-10 Layout for Project #5

Solder three flexible lengths (6 inches) of wire to the terminals. The projecting mounting tab can be broken off because it is not needed here. To identify the potentiometer leads, rotate the control fully clockwise and connect one lead of an ohmmeter across the center terminal. The other ohmmeter lead is touched to the outer terminals in turn. Mark the terminal that results in a 0-ohm reading. That terminal will be connected to the signal input. The remaining outer terminal is the ground terminal and should be marked as such.

Step 2

Prepare the rest of the mechanical components as follows. All components should have flexible lengths (6 inches) of extension wire in order to facilitate handling. Solder one terminal of the switch to the battery snap wire (red) and bring out leads from the other terminal and black battery snap wire. The jack socket and speaker should have similar pairs of leads. Identify and mark the positive and negative ends of the jack socket. The speaker leads can go either way.

Step 3

Locate the IC socket into the SINGMIN PCB as shown in Figure 17-10. Make sure that the notch faces left and the socket is seated flush with the board. Solder carefully, checking each pin in turn before moving to the next. Use a magnifier to visually check each solder joint thoroughly.

Step 4

Several solid wire links must be attached to set up the basic interconnection scheme for this circuit. Use the list below as a reference guide.

Link 1: Upper ground to lower ground
Link 2: Pin #3 to ground
Link 3: Pin #4 to ground
Link 4: Pin #6 to the positive supply

Step 5

Capacitor C3 (10 µF) is positioned to sit to the left of the socket and with the leads bent to fit into pin #1 and pin #8. The positive end of C3 must go to pin #1. You can choose to lay C3 either upright or on its side—there is plenty of space for either orientation. Solder it into place, check the solder integrity, and trim off the leads when the solder has cooled.

Step 6

Attach capacitor C2 (1,000 pF) from pin #2 to ground. C2 can be inserted either way. Take care not to damage the leads where they exit from the body.

Step 7

At pin #5 there is a resistor, R2 (10 ohm), and a series capacitor, C6 (0.1 µF), going to ground. R2 leads off from pin #5 to a floating terminal point, coupling C6 to ground. In addition, output capacitor C7 (100 µF) leads off to another takeoff point. The positive end of C7 goes to pin #5.

Step 8

Capacitor C4 (0.1 µF) goes from pin #6 to the positive supply. A second capacitor, C5 (100 µF), is also connected in parallel with C4, with the positive end of C5 going to pin #6.

Step 9

Add LED1 and resistor R3 (2.7 kohm) across the positive supply line and ground. The junction of LED1 and R3 terminates at an isolated junction.

That completes the first phase of the main electronic components. Before moving on to the next step, thoroughly check all of the connections for correct placement, orientation, and solder integrity.

Step 10

Add the switch S1 and battery snap combination between the positive supply line and ground.

Step 11

The input jack socket J1 is added to capacitor C1 and ground.

Step 12

The potentiometer R1 will have the three leads already identified. The terminal marked signal input goes to the free end of C1; the center terminal goes to pin #2; and the remaining lead is grounded.

Step 13

Insert the IC into the socket, aligning pin #1 with the socket's lower left-hand corner. Bend the IC leads carefully inward to fit the socket.

Step 14

Make sure the power switch S1 is in the off position and the volume control R1 is rotated fully counterclockwise. Attach a 9-volt battery. Momentarily switch it on. The LED should light and you should hear a click in the speaker. Rotate the volume control clockwise. All being well, there should be an increase in hiss level at the speaker. Touch the input signal pin of J1 with your finger. There should be an increase in hum level at the speaker. This is normal. If there is no sound at all, then switch it off and check all of your connections.

Conclusion

This volume-controlled amplifier makes a superb independent driver amplifier for testing many types of audio devices—from phonographs to tone generators to electric guitars. For greatest versatility, package all of the components in a plastic construction case and add labels to suit.

Project #6: Fixed-Gain Audio Preamplifier

The previous audio power amplifier project works best with a fairly large input signal. When trying to amplify a low-signal device, such as a microphone or a phonograph, some extra gain must be provided. This project's relatively simple circuit makes use of the general-purpose LM 741 operational amplifier, or op-amp. This op-amp normally needs a dual supply to operate—two 9-volt batteries; however, in keeping with the theme of this book to stay with a single 9-volt battery, we can use a few additional components to provide an artificial dual supply.

Circuit Description

Using the LM 741 op-amp begins with a simple design for an ac amplifier with a fixed gain of ×10. The op-amp, like any other amplifier, is a four-terminal device: one terminal for the positive supply, one for ground, one for the input, and one for the output. Because the design is for an audio preamplifier, there are the usual ac coupling capacitors at the input and output. The selection for the gain is made with just two resistors.

Parts List

Semiconductor
IC1: LM 741 operational amplifier

Resistors
R1 = 10 kohm
R2 = 10 kohm
R3 = 10 kohm
R4 = 100 kohm
R5 = 2.7 kohm

Capacitors
C1 = 0.1 µF
C2 = 100 µF
C3 = 0.1 µF
C4 = 0.1 µF
C5 = 100 µF

78 BEGINNING ANALOG ELECTRONICS THROUGH PROJECTS

Additional Parts and Materials
LED1: Light-emitting diode
J1: Miniature 1/8-inch jack socket
J2: Miniature 1/8-inch jack socket
S1: Miniature SPST toggle switch
B1: 9-volt battery
9-volt battery snap
8-pin IC socket
General purpose circuit assembly board
Hook-up wire (solid and stranded)

Pin Connections (see Figure 17-11)

Pin #1
There is no connection to this pin.

Pin #2
This pin goes to one end of the input resistor, R3 (10 kohm), and to one end of the feedback resistor, R4 (100 kohm). Signal gain is determined by the ratio of R4/R3. The input signal is fed via a capacitor, C1 (0.1 µF).

Pin #3
This pin goes to the artificial half-supply voltage, which is the junction of equal-value resistors R1 (10 kohm) and R2 (10 kohm). There is also a stabilizing capacitor, C2 (100 µF), attached to this midpoint and ground.

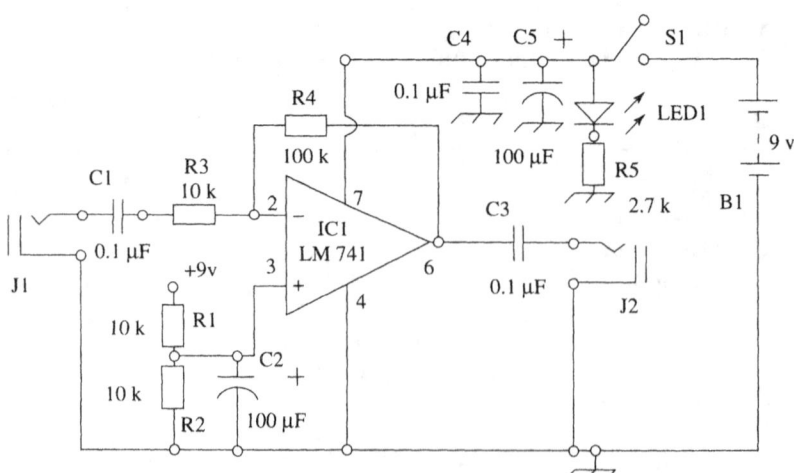

Figure 17-11   Project #6: Fixed-gain audio preamplifier

Pin #4
This pin goes to ground.

Pin #5
There is no connection to this pin.

Pin #6
The output signal exits from this pin via capacitor C3 (0.1 μF).

Pin #7
This pin goes to the positive supply voltage. Supply-stabilizing capacitors C4 (0.1 μF) and C5 (100 μF) run from pin #7 to ground.

Pin #8
There is no connection to this pin.

Light-emitting diode LED1 and current-limiting resistor R5 (2.7 kohm) are connected across the positive supply line and ground.

Connections to the outside world are made at the input through jack socket J1 and at the output from jack socket J2. Typically, a low-signal source—for example, a carbon microphone or a phonograph—would feed the input. The output would normally go to the input of an audio power amplifier, which in turn would drive a speaker.

Component Identification

Figure 17-11 shows the electrical schematic for the fixed-gain (×10) audio preamplifier. From the pin list above, identify all of the connections. Match all of the actual components.

Resistors
The resistor values should be matched as follows:

R1 = 10 kohm/color code = brown, black, orange
R2 = 10 kohm/color code = brown, black, orange
R3 = 10 kohm/color code = brown, black, orange
R4 = 100 kohm/color code = brown, black, yellow
R5 = 2.7 kohm/color code = red, violet, red

Capacitors
C1 = 0.1 μF/number code = 104'
C2 = 100 μF/electrolytic
C3 = 0.1 μF/number code = 104'
C4 = 0.1 μF/number code = 104'
C5 = 100 μF/electrolytic

Integrated Circuit

The LM 741 is an eight-pin device with the marking LM 741' on the surface. A circle stamped on the integrated circuit (IC) identifies pin #1. When correctly aligned, with the lettering the correct way up, the circle is at the lower left-hand corner. The notch in the IC always faces to the left.

Mechanical Components

The switch S1, battery snap, and 9-volt battery are the usual items needed to supply the power to the circuit. Additionally, there is a miniature 1/8-inch jack socket, J1, coupled to the input capacitor, C1. A similar jack socket, J2, is coupled to the output capacitor, C3.

Assembly Board

NOTE: The layout shown in Figure 17-12 is a suggestion only. There is plenty of space on the assembly board, and for beginners it is best to allow ample space between components. Because of differences in component sizes, feel free to vary the actual layout to suit yourself. Just make sure that the electrical connections are still as shown. Go through this diagram carefully, making sure that you follow each connection point. Match this diagram with the earlier electrical schematic in Figure 17-11.

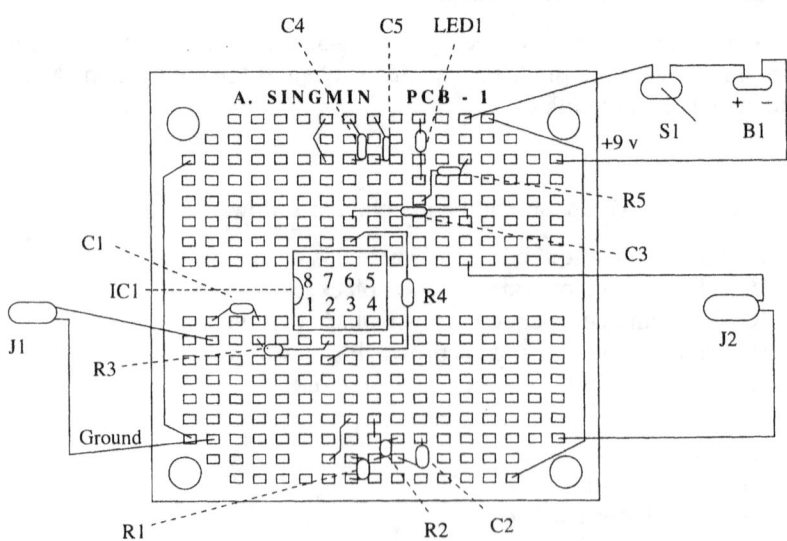

Figure 17-12 Layout for Project #6

Construction Details

Step 1
All of the mechanical components should have short (6 inches) flexible extension wires soldered to their terminals. Cover exposed solder connections with electrical tape to prevent accidental shorting. Switch S1 and battery snap are soldered as follows: one terminal of S1 to the red terminal of the battery snap wire. The completed assembly should then have a flying lead from the remaining switch terminal and another flying lead from the black (negative) battery snap wire.

The jack sockets have two flexible wires (6 inches) soldered to the signal and ground terminals. Make sure you clearly identify the signal and ground wires by labeling each wire.

Step 2
Place the IC socket into the SINGMIN PCB as shown in Figure 17-12, taking care that the notch faces left. Solder the socket terminals to the board, checking each step carefully before moving on to the next one.

Step 3
Several shorting links are needed as shown below:

Link 1: Upper ground to lower ground
Link 2: Upper positive supply to lower positive supply
Link 3: Pin #3 to half-supply point
Link 4: Pin #4 to ground
Link 5: Pin #7 to the positive supply

Step 4
Capacitor C1 ($0.1\,\mu F$) is soldered in first. Both ends go to floating terminals. One end will eventually go to the signal terminal of jack socket J1. The other end will go to resistor R3.

Step 5
Add resistor R3 (10 kohm) between C1 and pin #2.

Step 6
Add resistor R4 (100 kohm) between pin #2 and pin #6. Bend R4's leads to fit around the socket as shown.

Step 7
Solder the two equal-value resistors, R1 and R2, as shown. One end of R1 goes to the positive supply, and one end of R2 goes to the negative supply. The junction between R1 and R2 is coupled to pin #3 with the previous link. Because of the tight spacing, the resistor leads need to be bent as shown in Figure 17-12.

Step 8
Add capacitor C2 (100 µF) also between the resistor junction (R1 and R2) and ground. The negative end of C2 is grounded.

Step 9
Add the output capacitor C3 (0.1 µF) to pin #6. The other end of C3 terminates at a floating point.

Step 10
Along the top side of the IC socket, add capacitors C4 and C5. Note the polarity requirements for C5.

Step 11
Add the LED and resistor R5 (2.7 kohm) between the positive supply and ground.

Step 12
At this point, the first stage of construction is completed. Go thoroughly through all of the connections checking for errors. Pay particular attention to the integrity of solder joints.

Step 13
Finally, add the balance of mechanical components. Start with the switch/battery snap combo. Move next to the two jack sockets. They are both the same, so label them appropriately as input and output.

Step 14
Insert the LM 741 into the socket, making sure pin #1 is at the lower left-hand edge.

Step 15
Connect the preamplifier's output to the input of an audio power amplifier with an appropriate cable. Make sure the switch for the preamplifier is in the off position. Attach a 9-volt battery to the battery snap. Plug in a low-signal source to the preamplifier input, for example, a carbon microphone or phonograph. Switch on the power amplifier, keeping the volume control low to start with. Next, switch on the preamplifier. If you have a carbon microphone connected, there should be plenty of sound output from the speaker as you speak into the microphone. To check the effect of the preamplifier stage, switch off the preamp and then the power amp. Plug the microphone directly into the power amp, switch on, and now note that the volume is reduced.

Conclusion
Because there are only a few components, a small compact project case can be used to house all of the components. With labeling, the overall appearance can be quite attractive.

Project #7: Guitar Headphone Amplifier

This project revisits the popular LM 386 we have used previously. With a subtle variation in components, we now adapt the circuit to function as a guitar headphone amplifier. Rather than carrying your heavy guitar amplifier around with you all the time, use this lightweight unit as a practice amplifier. Late-night practicing can now be done without disturbing your neighbors.

Circuit Description

Amplification for an electric guitar is provided with the LM 386 audio power integrated circuit (IC). Two gain-setting components are added to boost the basic gain value from ×20 to ×50, giving it more than sufficient strength to drive a pair of headphones. The input is terminated with a 1/4-inch jack socket, to match the regular guitar cord/cable. The output terminates in a stereo 1/4-inch jack socket that is wired up as a mono socket because we will be using regular audio headphones for this project, like the type used with portable stereo cassette players. Because these headphones are stereo, the jack socket has to be a stereo type. The signal output from the amplifier, however, is a mono signal and, therefore, a slight bit of specialized wiring is needed when using the socket. This will be described later. Power is still supplied from a regular 9-volt battery.

Parts List

Semiconductor
IC1: LM 386 audio power amplifier

Resistors
R1 = 47 kohm
R2 = 10 kohm
R3 = 1 kohm
R4 = 10 ohm
R5 = 4.7 kohm

Capacitors
C1 = 0.47 µF
C2 = 1,000 pF
C3 = 10 µF
C4 = 0.1 µF
C5 = 100 µF
C6 = 0.1 µF
C7 = 100 µF

Additional Parts and Materials

LED1: Light-emitting diode
J1: Standard 1/4-inch jack socket
J2: Miniature 1/8-inch stereo jack socket
S1: Miniature SPST toggle switch
B1: 9-volt battery
8-pin IC socket
9-volt battery snap
General purpose circuit assembly board
Hook-up wire (solid and stranded)

Pin Connections (see Figure 17-13)

Pin #1

This is one of the gain-setting pins. A series capacitor, C3 (10 μF), and a resistor, R3 (1 kohm), tie pin #1 to pin #8. The values used boost the gain from the nominal ×20 to ×50 to provide more than enough amplification for the guitar output.

Pin #2

The signal input is fed to this pin. The signal is applied across resistor R1 (47 kohm), included to provide an appropriate match to the guitar output. The actual connection is made through a 1/4-inch mono jack socket J1. Because this is an audio amplifier, there is the usual capacitor coupling, C1 (0.47 μF). An

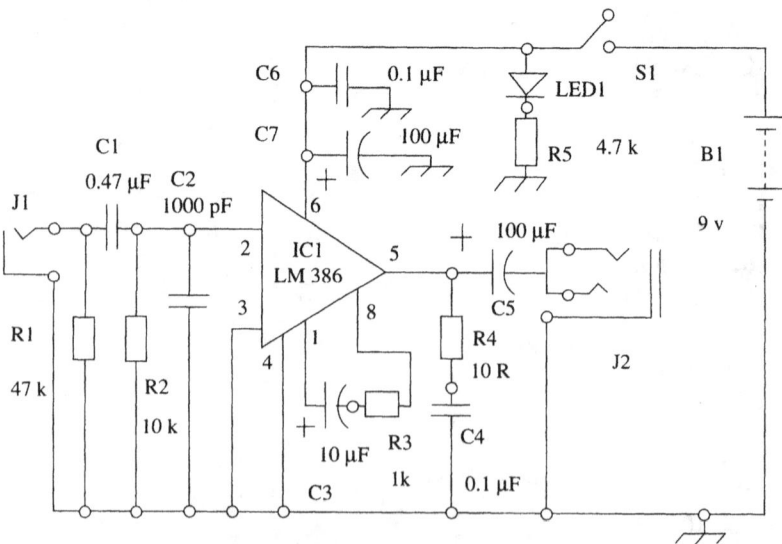

Figure 17-13 Project #7: Guitar headphone amplifier

additional resistor, R2 (10 kohm), works together with C1 to provide a more even response between the bass and treble notes. Capacitor C2 (1,000 pF) prevents any pickup of spurious radio signals.

Pin #3
This pin goes to ground.

Pin #4
This pin is the official ground pin.

Pin #5
The output signal comes from this pin. Because high-level signals will be passing through the amplifier, the Zobel network, resistor R4 (10 ohm) and capacitor C4 (0.1 µF), is added across pin #5 and ground. The actual signal output is taken through electrolytic capacitor C5 (100 µF). Finally, the real-world connection is from a miniature stereo jack socket, J2, wired up in the mono mode.

Pin #6
This pin is the positive supply connection. Capacitors C6 (0.1 µF) and C7 (100 µF) enhance the stability of the amplifier during operation.

Pin #7
There is no connection to this pin.

Pin #8
This is the other half of the gain-setting pins (pin #1 is the first half).

The light-emitting diode (LED1) and resistor R5 (4.7 kohm), wired across the positive supply and ground, complete the component count.

Component Identification

Figure 17-13 shows the electrical schematic for the guitar headphone amplifier. Use the pin connection listing above to identify and familiarize yourself with all of the components. Match the actual components.

Resistors
The resistor values used here should be matched using the color codes below:

R1 = 47 kohm/color code = orange, yellow, orange
R2 = 10 kohm/color code = brown, black, orange
R3 = 1 kohm/color code = brown, black, red
R4 = 10 ohm/color code = brown, black, black
R5 = 4.7 kohm/color code = orange, yellow, red

86 BEGINNING ANALOG ELECTRONICS THROUGH PROJECTS

Capacitors
C1 = 0.47 µF/number code = 474'
C2 = 1,000 pF/number code = 102'
C3 = 10 µF/electrolytic
C4 = 0.1 µF/number code = 104'
C5 = 100 µF/electrolytic
C6 = 0.1 µF/number code = 104'
C7 = 100 µF/electrolytic

Integrated Circuit

The LM 386 is an eight-pin device with the marking LM 386' on the top surface. Pin #1 is located at the lower left-hand edge near the identifying circle. The device wording reads the right way up when oriented correctly.

Mechanical Components

Switch S1, battery snap, and battery are the usual power-supplying components. Two new components are used here for the first time. The input jack socket, J1, is a 1/4-inch type, and the output socket, J2, is a miniature 1/8-inch stereo jack socket, wired in the mono mode.

Assembly Board

NOTE: The layout shown in Figure 17-14 is a suggestion only. There is plenty of space on the assembly board, and for beginners it is best to allow ample space between components. Because of differences in component sizes,

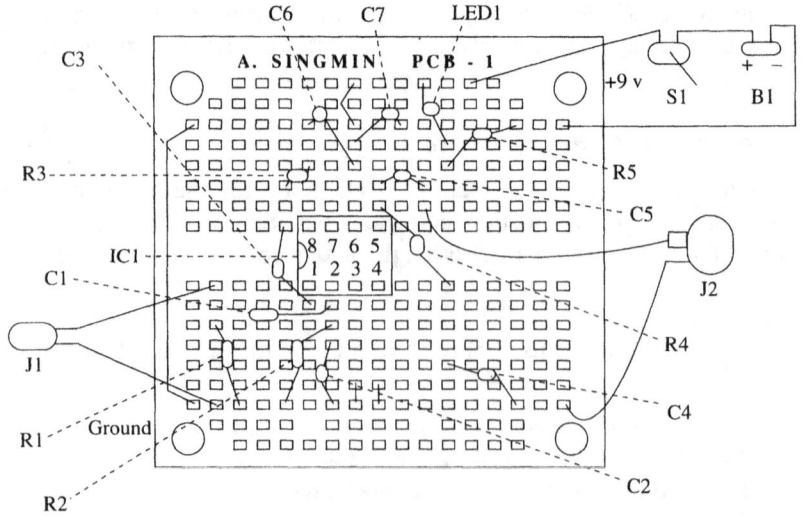

Figure 17-14 Layout for Project #7

feel free to vary the actual layout to suit yourself. Just make sure that the electrical connections are still as shown. Go through this diagram carefully, making sure that you follow each connection point. Match this diagram with the earlier electrical schematic in Figure 17-13.

Construction Details

Step 1

Connect one of the switch S1's terminals to the red lead of the battery snap. Use flexible wires (6 inches in length) between all connections. You should end up with two flying leads—one from the switch and one from the black battery clip lead. Insulate all bare solder connections.

Step 2

The two new jack sockets will require some detailed preparation before being usable. The input jack socket, J1, is a 1/4-inch mono type. In order to correctly identify the terminals, you will also need to procure a 1/4-inch jack plug that fits into J1. Unscrew the cover from the jack plug and use your ohmmeter to identify the tip and ground connections. The signal connection is made to the inner terminal and the ground connection to the outer terminal. Now insert the jack plug (with the cover still off) into the jack socket. There are two terminals on the jack socket. Use the ohmmeter again to identify the signal and ground connections. Solder flexible leads to the jack socket and mark the leads carefully as signal and ground.

For the output jack socket, J2, you will also need to have a stereo jack plug that fits into J2. Unscrew the cover from the barrel and note that the stereo jack plug has two signal connections and a ground. The ground terminal is the longer solder tag. Use the ohmmeter to verify the two signal connections. These correspond to the two furthermost tip connections at the shaft end. Plug the still-uncovered jack plug into the stereo jack socket. Use your ohmmeter to identify the ground connection on the jack socket. There are five separate connections on the stereo jack socket, so proceed carefully. This connection will be closest to where the jack plug enters the socket.

Next, identify the two jack socket terminals that correspond to the two signal input pins on the jack plug. Solder a shorting link on the jack socket between these two pins. You are ensuring that when a pair of stereo headphones is plugged into this stereo socket, fed by a mono signal, the sound will be heard in both earpieces. Finally, bring out two flexible leads from the stereo jack socket—one from ground and one from the linked connection. Mark the leads carefully as such.

Step 3

Locate the IC socket on the SINGMIN PCB as shown in Figure 17-14. Make sure the notch faces left and the socket is seated properly before soldering. Check each solder connection as you go along.

Step 4
Several wire links are needed here:

Link 1: Upper ground to lower ground
Link 2: Pin #3 to ground
Link 3: Pin #4 to ground
Link 4: Pin #6 to positive supply

Step 5
Gain-setting components R3 (1 kohm) and C3 (10 µF) are soldered first. Start with R3, bending the leads to conform to the shape shown. One end of R3 is attached to pin #8, and the other end terminates at a floating point. Capacitor C3 is an electrolytic component and sits vertically to the board. Note carefully the polarity. The positive end of C3 goes to pin #1.

Step 6
The shunting input resistor R1 (47 kohm) goes between ground and a floating point. Input capacitor C1 (0.47 µF) leads off from the floating end of R1 to terminate at pin #2. Resistor R2 (10 kohm) is added from pin #2 to ground. Capacitor C2 (1,000 pF) is also added between pin #2 and ground. Bend the leads accordingly to clear any adjacent components.

Step 7
A resistor, R4 (10 ohm), and a capacitor, C4 (0.1 µF), form the Zobel network for improving the large signal response. They are wired directly to the output, pin #5. Start with R4, taking one end to pin #5. The other end terminates at a floating point. Capacitor C4 joins this junction and completes the circuit to ground. The actual signal output is handled with electrolytic capacitor C5 (100 µF). The positive end goes to pin #5, with the negative end terminating in a floating point.

Step 8
Capacitors C6 (0.1 µF) and C7 (100 µF) run in parallel from pin #6 to ground. Capacitor C6, a disc type, presents no problem. Insert as shown. Correctly match the positive end of C7 to pin #6 and bend the leads slightly to clear adjacent components.

Step 9
The final electronic components are LED1 and resistor R5 (4.7 kohm). These are wired directly across the positive supply and ground. As with all series components, a free solder location is used as a floating point for the connection of the two components.

Step 10
Perform a thorough check now of all the components, looking for correctness of placements, lead orientations, and good solder joints.

Step 11

This next phase requires soldering all of the mechanical components. Start with the switch S1 and battery snap arrangement. Because this has already been prepared earlier, it is a simple process of soldering in two wires. Make sure the polarities are correct, with the black battery clip lead going to ground.

Step 12

The input jack socket J1 is attached to the floating end of R1. Take care with the polarity of J1, matching the correct ground and signal leads.

Step 13

The output jack socket J2 goes to the negative floating end of capacitor C5. Match the polarities correctly—ground to ground.

Step 14

Insert the IC carefully into the socket, aligning pin #1 with the lower left-hand corner of the socket.

Step 15

Plug in a pair of stereo headphones to J1. Make sure the switch is in the off position and attach a 9-volt battery. Place the headphones over your ears and momentarily switch it on. There should be a click and perhaps some hiss. Touch the signal input terminal with your finger. All being well, the sound should increase. If there is no sound, then switch it off and recheck all connections. Assuming all is well, though, plug in a guitar cord to the input, couple up a guitar, and switch it on. The sound output can be very loud, so use the guitar's volume control to adjust the sound.

Conclusion

A small, plastic project case makes an ideal finish for this project because the unit will be frequently transported with your guitar.

Project #8: Visual Electronic Metronome

In keeping with the musical theme of Project #7, this project offers a visual electronic metronome design using the familiar 555 timer chip. A conventional metronome produces a series of audible clicks that can be adjusted to match the pace of music being played. This project has a novel twist—the output is displayed on a single light-emitting diode (LED) instead. In this way, if you are recording, there is no unwanted metronome click on your tape. The component values have been chosen to produce beats that fit in the musical range, variable from around 35 to 350 beats per minute.

Circuit Description

The integrated circuit IC1 is a 555 timer used to produce a continuous stream of pulses. A combination of resistors and capacitors around pins #6 and #7 provides the frequency control. The number of beats per minute is adjusted with a potentiometer. Power is from a 9-volt battery. The output beat is displayed on a single LED.

Parts List

Semiconductors
IC1: LM 555 timer

Resistors
R1 = 1 kohm
R2 = 1 Mohm potentiometer
R3 = 1 kohm
R4 = 1 kohm

Capacitors
C1 = 2.2 µF
C2 = 0.01 µF
C3 = 0.1 µF
C4 = 100 µF

Additional Parts and Materials
LED1: Light-emitting diode
S1: Miniature SPST toggle switch
B1: 9-volt battery
Control knob to fit potentiometer
9-volt battery snap
8-pin IC socket
General purpose circuit assembly board
Hook-up wire (solid and stranded)

Pin Connections (see Figure 17-15)

Pin #1
This is the ground connection pin.

Pin #2
This pin is linked to pin #6 and is taken to ground with a timing capacitor, C1 (2.2 µF). There is also a timing resistor, R1 (1 kohm), running to pin #7.

Pin #3
The output comes from this pin. LED1 with its resistor R4 (1 kohm) ties this pin to ground.

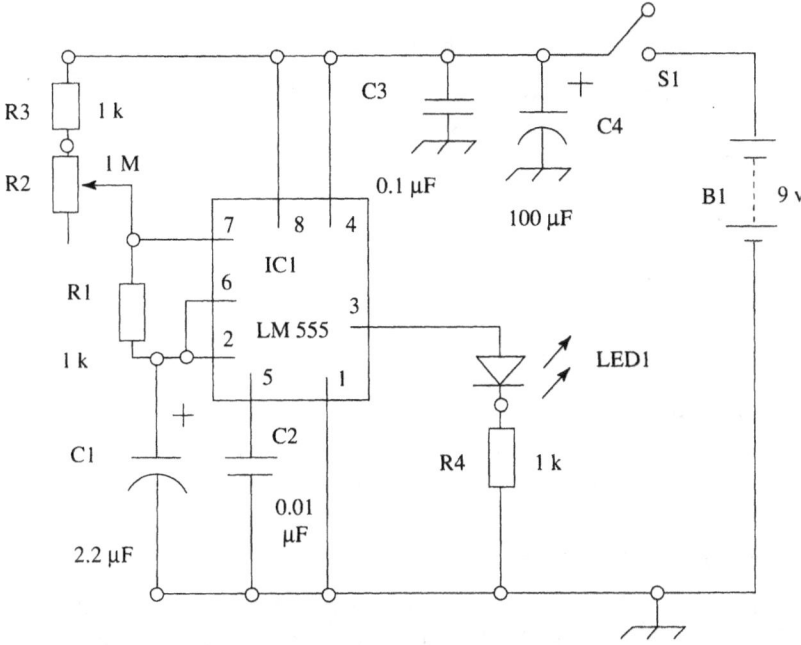

Figure 17-15 Project #8: Visual electronic metronome

Pin #4
This pin goes to the positive supply.

Pin #5
This pin goes via capacitor C2 (0.01 µF) to ground.

Pin #6
This pin is coupled via R1 (1 kohm) to pin #7.

Pin #7
This pin has a potentiometer, R2 (1 Mohm), and a series resistor, R3 (1 kohm), linking it to the positive supply rail.

Pin #8
This pin is connected to the positive supply rail.

A capacitor, C3 (0.1 µF), and an electrolytic capacitor, C4 (100 µF), across the supply voltage complete the circuit.

Component Identification

Figure 17-15 shows the electrical schematic for this project. Use the pin list above and carefully familiarize yourself with each component location. Next, match the actual components themselves with the schematic.

Resistors

Verify the resistor values with the color codes below:

R1 = 1 kohm/color code = brown, black, red
R3 = 1 kohm/color code = brown, black, red
R4 = 1 kohm/color code = brown, black, red

Capacitors

C1 = 2.2 µF/electrolytic
C2 = 0.01 µF/number code = 103'
C3 = 0.1 µF/number code = 104'
C4 = 100 µF/electrolytic

Integrated Circuit

This is an eight-pin pin device with the marking 555' on the top surface. A circle near the lower left-hand edge identifies pin #1.

Mechanical Components

The switch S1, battery snap, and 9-volt battery provide the supply requirements. Control of the frequency rate for the metronome is with a potentiometer, R2 (1 Mohm).

Assembly Board

NOTE: The layout shown in Figure 17-16 is a suggestion only. There is plenty of space on the assembly board, and for beginners it is best to allow

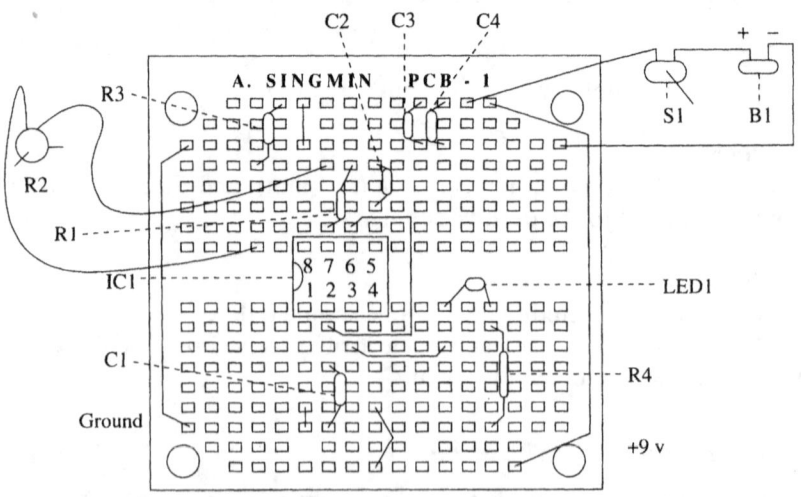

Figure 17-16 Layout for Project #8

ample space between components. Because of differences in component sizes, feel free to vary the actual layout to suit yourself. Just make sure that the electrical connections are still as shown. Go through this diagram carefully, making sure that you follow each connection point. Match this diagram with the earlier electrical schematic in Figure 17-15.

Construction Details

Step 1
The switch S1 and battery snap are linked together, with one switch terminal going to the red battery snap lead. Flexible extension wires (6 inches in length) are used to extend the remaining connections, forming the familiar power supply combination.

The potentiometer requires that the shaft be measured to fit the knob and then cut to size. Remove any sharp edges. Turn the potentiometer shaft fully clockwise and use your ohmmeter to verify which of the outer terminals is short-circuited to the center terminal. That will be the terminal to use. Mark it as needed. Solder two flexible leads out from the center and this terminal.

Step 2
Place the IC socket in position on the board and solder into position. Check each lead carefully for solder integrity as you go along. The socket should sit square and flush with the board when completed, with shiny solder joints around each terminal.

Step 3
A few solid wire links need to be installed next. Use the list below as a guide:

Link 1: Upper ground to lower ground
Link 2: Upper positive supply to lower positive supply
Link 3: Pin #1 to ground
Link 4: Pin #2 to pin #6
Link 5: Pin #4 to the positive supply
Link 6: Pin #8 to the positive supply

Step 4
Install capacitor C1 ($2.2\,\mu F$) between pin #2 and ground. The negative end goes to ground. The device sits upright.

Step 5
Resistor R1 (1 kohm) goes in next between pin #6 and pin #7. The space is limited, so the resistor is installed in a vertical fashion.

Step 6

Resistor R3 (1 kohm) is first soldered from the positive supply line (near pin #7) and a floating point. R3 is placed flush with the board because there is plenty of room. Later, the connection to R3 and pin #7 will be made through potentiometer R2 (1 Mohm).

Step 7

LED1 and resistor R4 (1 kohm) are added across pin #3 and ground. The junction between LED1 and R4 goes to a floating point.

Step 8

Check all of the connections carefully before proceeding to the next stage.

Step 9

The few mechanical components go in next. Start with the switch/battery snap combo. Solder the two flying leads into the board as shown. Take care with the polarity. The black battery snap lead goes to ground, and the switch terminal goes to the positive supply line.

Step 10

Insert the 555 timer IC carefully into the socket. Gently bend the leads inward if required. Check that pin #1 on the IC lines up with pin #1 on the socket. The legend on the IC will read the correct way up.

Step 11

Make sure the switch is in the off position. Insert a 9-volt battery into the battery clip. Turn the potentiometer fully counterclockwise. Momentarily switch it on. The LED will start to pulse if all is well. Rotate the potentiometer and the pulse rate will increase. Switch it off if there is no sign of life and recheck all connections.

Conclusion

This metronome makes an ideal timekeeper for practice sessions. Mount all of the components in a small project case for convenience.

Project #9: Variable-Gain, Hi/Lo Response Audio Preamplifier

In Project #6, we built a simple fixed-gain (×10) audio preamplifier. Here we increase the versatility of that basic circuit by making the gain continuously variable from ×1 to around ×100, which will cover most application requirements. This is still an audio preamplifier, so the customary capacitors are at the input and output.

Circuit Description

The LM 741 integrated circuit (IC) is again used for this project. This circuit is based on that used in Project #6, except for a change from a fixed-feedback resistor to a series variable-and-fixed resistor combination. Also included in this design is a high-frequency limiting capacitor that is controlled by a switch and that will give you a choice between either a sharp or a mellow sound. Even though the description is similar to Project #6, I suggest you build a new and separate circuit for this project and do not try to modify the circuit already constructed in Project #6.

Parts List

Semiconductor
IC1: LM 741 operational amplifier

Resistors
R1 = 10 kohm
R2 = 10 kohm
R3 = 10 kohm
R4 = 10 kohm
R5 = 1 Mohm potentiometer
R6 = 4.7 kohm

Capacitors
C1 = 0.1 µF
C2 = 100 µF
C3 = 0.0047 µF
C4 = 0.1 µF
C5 = 0.1 µF
C6 = 100 µF

Additional Parts and Materials
LED1: Light-emitting diode
J1: Miniature 1/8-inch jack socket
J2: Miniature 1/8-inch jack socket
S1: Miniature SPST toggle switch
S2: Miniature SPST toggle switch
B1: 9-volt battery
9-volt battery snap
8-pin IC socket
General purpose circuit assembly board
Hook-up wire (solid and stranded)

Pin Connections (see Figure 17-17)

Pin #1

There is no connection to this pin.

Pin #2

This pin goes to one end of the input resistor, R1 (10 kohm), and to one end of the feedback resistor combination, which consists of a fixed resistor, R4 (10 kohm), in series with a variable resistor, R5 (1 Mohm). Signal gain is thus determined by the ratio of (R4 + R5) ÷ R1. The gain can be increased by decreasing R5. A further capacitor, C3 (0.0047 µF), is connected via a switch, S1, across the R4 + R5 combination. The purpose of C3 is to limit the high-frequency response of the circuit. From a practical viewpoint, the sound will be mellower when C3 is switched in and sharper when left out. This addition gives you a bit of tonal variation. At the input, the signal is fed via a capacitor, C1 (0.1 µF).

Pin #3

This pin goes to the artificial half-supply voltage, which is the junction of equal-value resistors R2 (10 kohm) and R3 (10 kohm). There is also a stabilizing capacitor, C2 (100 µF), which is attached to this midpoint and ground.

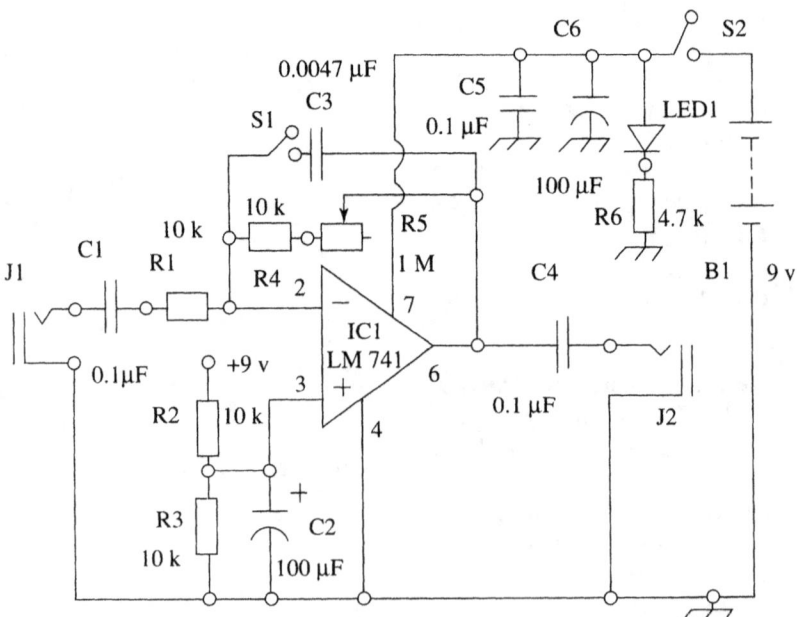

Figure 17-17 Project #9: Variable-gain, hi/lo response audio preamplifier

Construction Details for 10 Simple Projects

Pin #4
This pin goes to ground.

Pin #5
There is no connection to this pin.

Pin #6
The output signal exits from this pin via capacitor C4 (0.1 μF).

Pin #7
This pin goes to the positive supply voltage. Supply stabilizing capacitors, C5 (0.1 μF) and C6 (100 μF), run from pin #7 to ground.

Pin #8
There is no connection to this pin.

Light-emitting diode LED1 and current-limiting resistor R6 (4.7 kohm) are connected across the positive supply line and ground.

Connections to the outside world are made at the input through jack socket J1 and at the output from jack socket J2. A variety of signal sources can be accepted because the stage gain is continuously variable. The output would normally go to the input of an audio power amplifier that, in turn, would drive a speaker.

Component Identification

Figure 17-17 shows the electrical schematic for the variable-gain, hi/lo response audio preamplifier. From the pin list above, identify all of the connections. Match all of the actual components.

Resistors
The resistor values should be matched as follows:

R1 = 10 kohm/color code = brown, black, orange
R2 = 10 kohm/color code = brown, black, orange
R3 = 10 kohm/color code = brown, black, orange
R4 = 10 kohm/color code = brown, black, orange
R6 = 4.7 kohm/color code = orange, violet, red

Capacitors
C1 = 0.1 μF/number code = 104'
C2 = 100 μF/electrolytic
C3 = 0.0047 μF/number code = 472'
C4 = 0.1 μF/number code = 104'
C5 = 0.1 μF/number code = 104'
C6 = 100 μF/electrolytic

Integrated Circuit

The LM 741 is an eight-pin device with the marking LM 741' on the surface. A circle stamped on the IC identifies pin #1. When correctly aligned, with the lettering the correct way up, the circle is at the lower left-hand corner. The notch in the IC always faces to the left.

Mechanical Components

A miniature on/off switch, S1, allows capacitor C3 to be switched out or in, thus providing the high or low response. The power switch S2, battery snap, and 9-volt battery are the usual items needed to supply the power to the circuit. Additionally, there is a miniature 1/8-inch jack socket, J1, coupled to the input capacitor, C1. A similar jack socket, J2, is coupled to the output capacitor, C4.

Assembly Board

NOTE: The layout shown in Figure 17-18 is a suggestion only. There is plenty of space on the assembly board, and for beginners it is best to allow ample space between components. Because of differences in component sizes, feel free to vary the actual layout to suit yourself. Just make sure that the electrical connections are still as shown. Go through this diagram carefully, making sure that you follow each connection point. Match this diagram with the earlier electrical schematic in Figure 17-17.

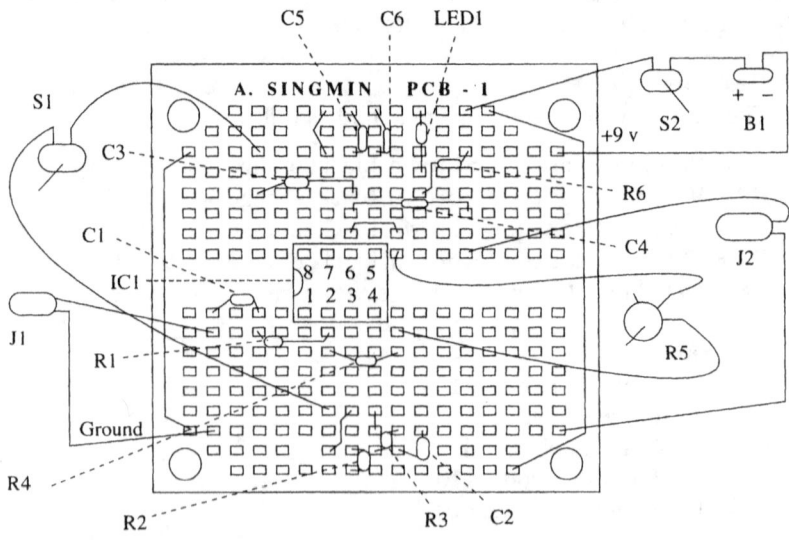

Figure 17-18 Layout for Project #9

Construction Details

Step 1
All of the mechanical components should have short (6 inches), flexible extension wires soldered to their terminals. Add two short flexible leads to switch S1, one from the center terminal and the other from either of the remaining terminals. Cover exposed solder connections with electrical tape to prevent accidental shorting. The switch S2 and battery snap are soldered as follows: One terminal of S2 goes to the red terminal of the battery snap. The completed assembly should then have a flying lead from the remaining switch terminal and another flying lead from the black (negative) battery clip.

The jack sockets J1 and J2 have two flexible wires (6 inches in length) soldered to the signal and ground terminals. Make sure you clearly identify the signal and ground wires by labeling each wire.

Step 2
Place the IC socket into the SINGMIN PCB as shown in Figure 17-18, taking care that the notch faces left. Solder the socket terminals to the board, checking each step carefully as you go along before moving on to the next step.

Step 3
Several shorting links are needed as shown below:

Link 1: Upper ground to lower ground
Link 2: Upper positive supply to lower positive supply
Link 3: Pin #3 to half-supply point
Link 4: Pin #4 to ground
Link 5: Pin #7 to the positive supply

Step 4
Begin with resistor R1 (10 kohm) soldered between pin #2 and a floating point.

Step 5
Add capacitor C1 (0.1 µF) between the floating point above and another floating point. This will be for the signal input components.

Step 6
Add resistor R4 (10 kohm) between pin #2 and a floating point. This will be for the feedback gain function.

Step 7
Add capacitor C3 (0.0047 µF) between pin #6 and a floating terminal. This will be for the hi/lo response function.

Step 8
Capacitor C1 (0.1 µF) is added to the floating end of R1 (10 kohm). Its free end is tied to a floating point. This will eventually go to the input jack socket J1.

Step 9
Solder the two equal resistors, R2 and R3, as shown. One end of R2 goes to the positive supply, and one end of R3 goes to the negative supply. The junction between R2 and R3 is coupled to pin #3 with the previous link. Because of the tight spacing, the resistor leads need to be bent as shown in Figure 17-18.

Step 10
Add capacitor C2 (100 µF) also between the resistor junction (R2 and R3) and ground. The negative end of C2 is grounded.

Step 11
Add the output capacitor C4 (0.1 µF) to pin #6. The other end of C4 terminates at a floating point.

Step 12
Along the top side of the IC socket, add capacitors C5 (0.1 µF) and C6 (100 µF) from pin #7 to the positive supply. Note the polarity requirements for C6.

Step 13
Add the LED and resistor R6 (4.7 kohm) between the positive supply and ground.

Step 14
The first stage of construction is now completed. Go thoroughly through all of the connections, checking for errors. Pay particular attention to the integrity of solder joints.

Step 15
Finally, add the balance of mechanical components. Start with the switch S1. One end goes from the free end of capacitor C3 and pin #2. Next, add the power switch S2 and battery snap combo. Move next to the two jack sockets, J1 and J2. They are both the same, so label them as input and output.

Step 16
Insert the LM 741 into the socket, making sure pin #1 is at the lower left-hand edge.

Step 17
Connect the preamplifier's output to the input of an audio power amplifier with an appropriate cable. Make sure the switch for the preamplifier

is in the off position. Attach a 9-volt battery to the snap. Turn the gain-control potentiometer R5 to a midpoint setting. Put the switch S1 in the off position. Plug in a low-signal source to the preamplifier input, for example, a carbon microphone or a phonograph. Switch on the power amplifier, keeping its volume control low to start with. Next switch on the preamplifier. If you have a carbon microphone connected, there should be plenty of sound output from the speaker as you speak into the microphone. Rotate R5 clockwise and the sound will increase. The opposite effect should occur for a counterclockwise rotation. Switch S1 can be toggled for either a high or a low response.

Conclusion

Choose a spacious project box to fit all of the components and add labels to suit.

Project #10: Dual-Gain Electret Microphone Audio Preamplifier

For hobby applications, the electret microphone element is an ideal update to the older carbon microphone. The electret microphone element is compact, sensitive, and easy to use; however, unlike the carbon microphone that can be plugged directly into a general-purpose amplifier or preamplifier, the electret requires a dc bias voltage before it can operate correctly. This is not a problem to implement and requires just three extra resistors.

Our final project, therefore, describes a dual-gain (select ×10 or ×100) electret microphone preamplifier. There is also a switch included to choose whether you want the dc bias or not. In this way, the unit can handle both electret or carbon microphones.

Circuit Description

Our old favorite op-amp, the LM 741, is again used. By now, if you have been building the previous projects, this one should be relatively easy. Gain is set by the ratio of the feedback to input resistors. There are two feedback resistors, each of which can be selected with a switch to give the desired gain. A second switch selects or ignores the dc bias voltage, so the project can be used with either type of microphone. The usual coupling capacitors feed the input and output. A split dc bias supply voltage provides the correct bias to the non-inverting pin.

Parts List

Semiconductor
IC1: LM 741 operational amplifier

Resistors
R1 = 1 kohm
R2 = 2.7 kohm
R3 = 2.7 kohm
R4 = 10 kohm
R5 = 10 kohm
R6 = 10 kohm
R7 = 100 kohm
R8 = 1 Mohm
R9 = 4.7 kohm

Capacitors
C1 = 0.1 µF
C2 = 100 µF
C3 = 0.1 µF
C4 = 0.1 µF
C5 = 100 µF

Additional Parts and Materials
LED1: Light-emitting diode
J1: Miniature 1/8-inch jack socket
J2: Miniature 1/8-inch jack socket
S1: Miniature SPST toggle switch
S2: Miniature SPDT toggle switch
S3: Miniature SPST toggle switch
MC1: electret microphone element
B1: 9-volt battery
9-volt battery snap
8-pin IC socket
General purpose circuit assembly board
Hook-up wire (solid and stranded)

Pin Connections (see Figure 17-19)

Pin #1
There is no connection to this pin.

Pin #2
This pin goes to one end of the input resistor R6 (10 kohm) and eventually to one end of the dual-feedback resistor combination, R7 (100 kohm) and R8 (1 Mohm). A switch, S2, selects either R7 or R8. Signal gain is thus determined by the ratio of R7/R6 or R8/R6, giving a gain of ×10 or ×100. At the input end, the signal is fed via a capacitor, C1 (0.1 µF).

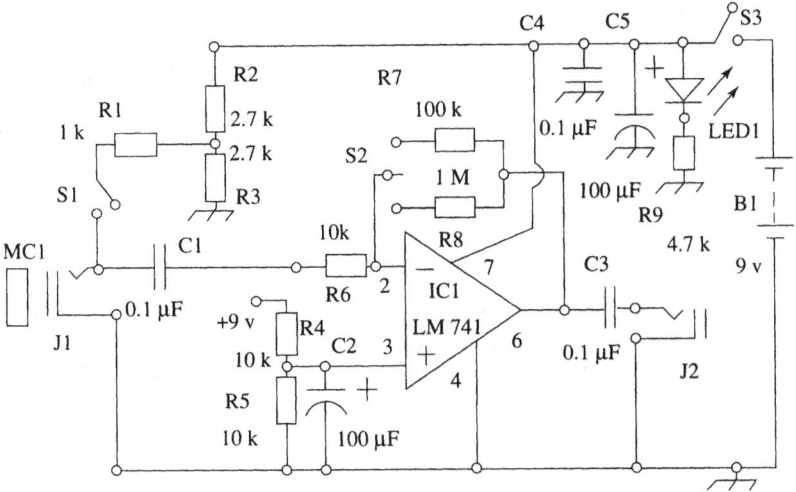

Figure 17-19 Project #10: Dual-gain electret microphone preamplifier

Pin #3
This pin goes to the artificial half-supply voltage, which is the junction of equal-value resistors R4 (10 kohm) and R5 (10 kohm). There is also a stabilizing capacitor, C2 (100 µF), attached to this midpoint and ground.

Pin #4
This pin goes to ground.

Pin #5
There is no connection to this pin.

Pin #6
The output signal exits from this pin via capacitor C3 (0.1 µF).

Pin #7
This pin goes to the positive supply voltage. Supply-stabilizing capacitors C4 (0.1 µF) and C5 (100 µF) run from pin #7 to ground.

Pin #8
There is no connection to this pin.

Two resistors, R2 (2.7 kohm) and R3 (2.7 kohm), are connected across the positive supply and ground. This produces a half-supply voltage point. The junction is fed through resistor R1 (1 kohm). The other end of R1 goes to a switch, S1, that in turn is coupled to the signal end of the microphone. Closing the switch will thus apply the necessary bias for an electret microphone.

Light-emitting diode (LED1) and current-limiting resistor R9 (4.7 kohm) are connected across the positive supply line and ground.

Connections to the outside world are made at the input through jack socket J1 and at the output from jack socket J2. The output would normally go to the input of an audio power amplifier, which in turn would drive a speaker. Power is fed through switch S3.

Component Identification

Figure 17-19 shows the electrical schematic for the electret microphone, dc bias, dual-gain (×10 and ×100) audio preamplifier. From the pin list above, identify all of the connections. Match up all of the actual components.

Resistors
The resistor values should be matched as follows:

R1 = 1 kohm/color code = brown, black, red
R2 = 2.7 kohm/color code = red, violet, red
R3 = 2.7 kohm/color code = red, violet, red
R4 = 10 kohm/color code = brown, black, orange
R5 = 10 kohm/color code = brown, black, orange
R6 = 10 kohm/color code = brown, black, orange
R7 = 100 kohm/color code = brown, black, yellow
R8 = 1 Mohm/color code = brown, black, green
R9 = 4.7 kohm/color code = yellow, violet, red

Capacitors
C1 = 0.1 µF/number code = 104'
C2 = 100 µF/electrolytic
C3 = 0.1 µF/number code = 104'
C4 = 0.1 µF/number code = 104'
C5 = 100 µF/electrolytic

Integrated Circuit
The LM 741 is an eight-pin device with the marking LM 741' on the surface. A circle stamped on the integrated circuit (IC) identifies pin #1. When correctly aligned, with the lettering the correct way up, the circle is at the lower left-hand corner. The notch in the IC always faces to the left.

Mechanical Components

A miniature single pole, double throw (SPDT) switch, S2, selects either resistor R7 or R8. A miniature on/off switch, S1, switches the microphone bias supply in or out. Power switch S3, battery snap, and a 9-volt battery are the usual items needed to supply the power to the circuit. Additionally, there is a miniature 1/8-inch jack socket, J1, coupled to the input capacitor C1. A similar jack socket, J2, is coupled to the output capacitor C3.

Assembly Board

NOTE: The layout shown in Figure 17-20 is a suggestion only. There is plenty of space on the assembly board, and for beginners it is best to allow ample space between components. Because of differences in component sizes, feel free to vary the actual layout to suit yourself. Just make sure that the electrical connections are still as shown. Go through this diagram carefully, making sure that you follow each connection point. Match this diagram with the earlier electrical schematic in Figure 17-19.

Construction Details

Step 1

All of the mechanical components should have short (6 inches) flexible extension wires soldered to their terminals. Add three short flexible leads to switch S2. Cover exposed solder connections with electrical tape to prevent accidental shorting. For switch S1, use just two terminals, the center and either of the remaining ones. The switch S3 and battery snap are soldered as follows: one terminal of S3 to the red terminal of the battery snap. The completed assembly should then have a flying lead from the remaining switch terminal and another flying lead from the black (negative) battery snap.

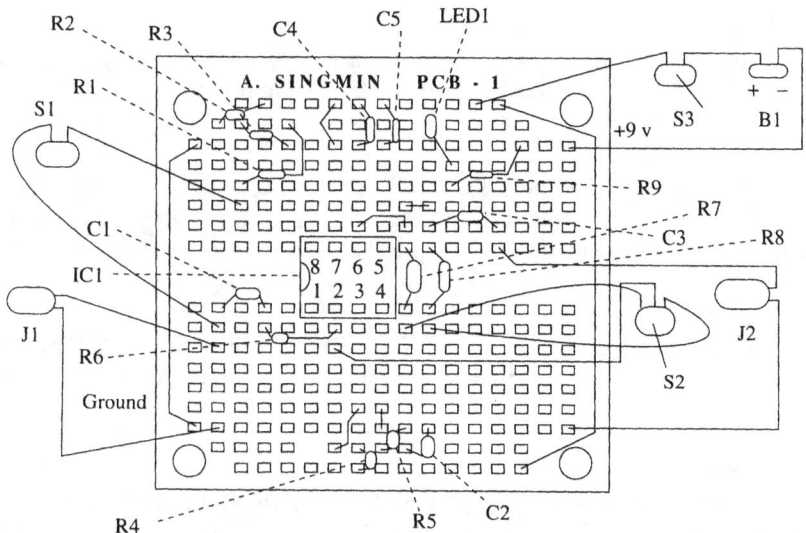

Figure 17-20 Layout for Project #10

The jack sockets J1 and J2 have two flexible wires (6 inches) soldered to the signal and ground terminals. Make sure you clearly identify the signal and ground wires by labeling each wire.

Step 2
Place the IC socket into the SINGMIN PCB as shown in Figure 17-20, taking care that the notch faces left. Solder the socket terminals to the board, checking each step carefully as you go along before moving to the next.

Step 3
Several shorting links are needed as shown below:

Link 1: Upper ground to lower ground
Link 2: Upper positive supply to lower positive supply
Link 3: Pin #3 to half-supply point
Link 4: Pin #4 to ground
Link 5: Pin #7 to the positive supply

Step 4
Begin with resistor R6 (10 kohm) soldered between pin #2 and a floating point.

Step 5
Add capacitor C1 (0.1 µF) between the floating point above. The other end goes to another floating point that will eventually terminate in the input jack socket J1.

Step 6
Add resistor R7 (100 kohm) between pin #6 and a floating point. Add resistor R8 (1 Mohm) between pin #6 and a floating point.

Step 7
Attach the two resistors R4 (10 kohm) and R5 (10 kohm) between the positive and ground rails as shown. Capacitor C2 (100 µF) goes from the junction of R4 and R5 to ground. Make sure the negative end of C2 is grounded.

Step 8
Attach one end of capacitor C3 (0.1 µF) to the output pin #6. The other end goes to a floating point that will eventually terminate in the output jack socket J2.

Step 9
Solder capacitors C4 (0.1 µF) and C5 (100 µF) between pin #7 and ground. C5 is polarity sensitive and must have the negative end connected to ground.

Step 10
Bias resistors R2 (2.7 kohm) and R3 (2.7 kohm) are added across the positive supply and floating point and across ground and the same floating point. One end of resistor R1 (1 kohm) goes to this same floating junction. The other end of R1 terminates in a new floating junction that will eventually go to the selector switch S1.

Step 11
Add the light-emitting diode LED1 and resistor R9 (4.7 kohm) between the positive supply and ground.

Step 12
The first stage of construction is now completed. Go thoroughly through all of the connections, checking for errors. Pay particular attention to the integrity of solder joints.

Step 13
Finally, add the balance of mechanical components. Start with the switch S2. The center terminal goes to pin #2. The other two ends go to the free ends of R7 and R8. It doesn't matter which way around it goes.

Step 14
The next switch, S1, goes between the free end of R1 and the free end of C1.

Step 15
The input jack socket J1 goes to this floating end of C1. The signal terminal goes to C1, and the ground terminal goes to ground. Label this socket as the input socket.

Step 16
Add output jack socket J2 to the floating end of C3. The signal terminal goes to C3. The ground terminal goes to ground. Label this socket as the output socket.

Step 17
Finally, add the power switch S3 and battery snap combo.

Step 18
Insert the LM 741 into the socket, making sure pin #1 is at the lower left-hand edge.

Step 19
Connect the preamplifier's output to the input of an audio power amplifier with an appropriate cable. Set switch S2 to the ×10 gain position and switch

the dc bias voltage on. Connect an electret microphone element to a 1/8-inch jack plug and plug into the input socket J1. Make sure the switch for the pre-amplifier is in the off position. Attach a 9-volt battery to the snap. Switch on the power amplifier, keeping its volume control low to start with. Next switch on the preamplifier. There should be plenty of sensitivity from the microphone. Flip switch S1 to the ×100 position. The sound will dramatically increase. Turn the main amplifier's volume down if the level is too high. Switch off S1. Notice that the microphone now does not function. Substituting a carbon microphone at this point, which should function with this setting, is another way to check this circuit.

Conclusion

As before, choose a spacious project box to fit all of the components.

CHAPTER **18**

Troubleshooting Test Equipment

These two new projects are included in the second edition to give you an opportunity to express your skills in devising your own board layout designs. The circuit schematics are straightforward, and together they provide you with a pair of useful troubleshooting pieces of equipment. Two of the most useful circuits for troubleshooting a nonfunctioning audio stage are a signal injector and a signal monitor. The signal injector is an audio signal source, which is compact and easy to use, and which generates a fixed square wave frequency in the audio frequency band. By injecting this tracer signal, you can systematically verify the signal path through to the output of your project under test and by verifying if you get an output signal. If there is one present, then this means that particular stage under test is functioning correctly.

The signal monitor is the complementary circuit to the signal injector. It enables you to detect the presence of an injected signal going through the amplifier under test. Both test units operate in the audio frequency band. There is no circuit layout because this is left up to your imagination and resourcefulness.

Project #11: Signal Injector

Introduction

Many simple audio circuits, such as the type described in this book, fall quite simply into two categories: preamplifiers and power amplifiers. As you progress with building proficiency, you'll probably start to cascade various audio building blocks together. For example, this cascade chain could be a two-stage preamplifier or a preamplifier followed by a filter circuit. If you find that the total circuit when completed is not functioning, then the principle of testing is to check out each stage in turn.

That's the benefit of the signal injector. It allows you to check each stage in turn, starting first with the last stage and working backward. Because it is designed as a tester circuit, the signal injector circuit has all unnecessary components removed, and hence it becomes compact, so much so that you'll be able to house it in a very small case. Typically, I've found that plastic pill containers, especially the larger sizes, make useful housings. Choose a size that comfortably accommodates a 9-volt battery, plus some additional space to take a small section of assembly board.

Circuit Description

Figure 18-1 shows the electrical schematic for the signal injector. The principal focus of this signal injector design is to reduce the component count to an absolutely bare minimum, in order to get all of the components into as small a housing as possible. We're going to include only components that are absolutely necessary. That's the innovative part of this design (and the next one)—it's an emphasis more on how the circuit's put together rather than on the electronic circuitry itself. A seemingly simple and well-used circuit can take on the guise of a totally new and unexpected appearance, providing yet another slant on an electronic innovation. This circuit is essentially a signal generator operating at a fixed (audio) frequency, but it's how you use it that counts!

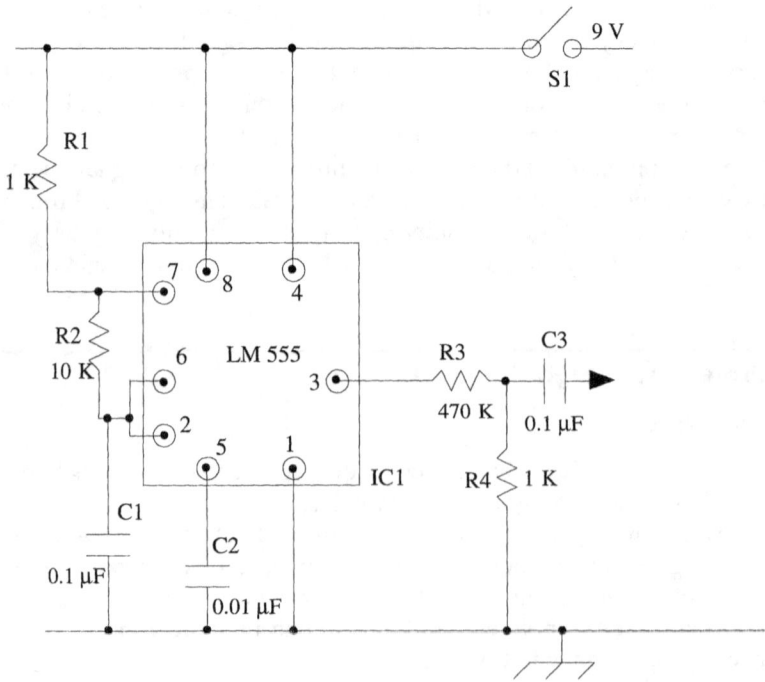

Figure 18-1 Signal injector

The LM 555 timer (IC1) is used as the principal device for generating a fixed test frequency in the audio frequency band. It's nice and easy to use, as you've no doubt seen before in other circuits. When configured to work as a free-running oscillator (in the stable mode), we see that only seven components (not counting the mechanical components) are required in this design: one IC, four resistors, and three capacitors—that's the minimum we can get away with. The timing components are resistor R1 (1 kohm) running from pin #7 to Vcc, resistor R2 (10 kohm) coupled from pin #6 to pin #7, and capacitor C1 (0.1 µF) taken from pin #6 to ground. Pin #2 is coupled to pin #6, as is always the case with this device when it is designed to generate a continuous train of pulses.

Pin #5 has a capacitor C2 (0.01 µF) connected to ground. The signal output from the LM 555 is a square wave, which is quite acceptable for the purposes we're designing for. The magnitude of the output signal will be very large, close to the supply voltage, in fact. This is far too large to be of use as a test signal. In order to provide some attenuation, two resistors, R3 (470 kohm) and R4 (1 kohm), reduce the signal down to about 20 millivolts, which is a much more suitable level. The output from pin #3 is capacitively coupled out via C3 (0.1 µF).

Power connections to IC1 are to pin #8 and pin #4. Pin #1 is the ground connection. The power feed from a 9-volt battery is via switch S1. As the battery voltage drops, the output will decrease in amplitude, too, but this is unimportant in this application. The customary power supply components have been shorn off in this minimalist design.

Construction Details

Here's where the really innovative work begins. We're looking to create as small an assembly board arrangement as possible. The size of the required board depends on the size of container you choose to house the component. Let's start with what the overall form will be. The basic premise is to use a pill container. Any large size will do. Measure a regular 9-volt battery, and gauge the container size needed from there. Allow extra headroom for the board assembly, which we'll get to soon.

Once the container is selected, we're going to have a solid probe tip protruding from one end and a flexible flying ground lead emerging from the opposite end. Somewhere along the body part will be mounted a small on/off switch—the smallest you can find. The best type to use, from the point of taking up the least space, is the type that's designed to be soldered directly onto a PCB. We'll be attaching this switch directly to the container with an epoxy glue.

Drill a small hole carefully into the body of the container when the board has been assembled and you can position the switch to clear the components. All of the project sections should be assembled outside of the container first to make sure it works and everything fits properly and with adequate clearance.

The flexible type of container is preferred over the hard plastic type. Drilling holes into the flexible type is easier.

The probe tip needs to have a thin, pointed end, like the test leads supplied with a multimeter. You can usually purchase the multimeter leads separately and just use one of them. Cut the test lead to fit, leaving a short but sufficient section of flexible wire to later join to the assembly board. The remaining half of the test lead will be used later for the flying ground lead, so use the black lead (rather than the red lead) because black will coincide with the fact that it is a ground lead. The probe tip is epoxied onto the end cap.

As a nice touch, a section of red sleeving can be placed over the black test lead to indicate that it's the "live" end of the signal injector. The ground lead exits through the base of the container and needs to be terminated with a small alligator clip because it will be clipped to a ground terminal of the project under test. That's the switch, test tip, and ground wire sorted out.

Next, we come to the board assembly itself. First, estimate the size of the board needed to fit the container, and then verify that all of the components will fit on it. Use an IC socket to house IC1. The four resistors (R1, R2, R3, and R4) are laid down flat on the board and won't take up any space. The three capacitors (C1, C2, and C3) are also laid down in a like manner on the board.

That's all of the components taken care of. The rest of the connections are simple wire connections from pin #6 to pin #2, the power connections from pin #8 and pin #4, and the ground connection to pin #1. The flying ground lead can be taken from any convenient point on the board. The probe tip is taken from capacitor C3. Wire up a 9-volt battery snap to the switch. You should be able to easily fit these few components onto a section of board that has the same outline as the 9-volt battery.

Before committing the sections to permanent mounting, check out that the system operates correctly first. The circuit is fairly simple as it is, but it is going to be a very frustrating exercise to have to disassemble everything if you haven't done the verification beforehand. Use multistranded wire for any connection that's going to be flexed during the assembly process—the thinner the gauge the better, to conserve space. Use an oscilloscope or an audio amplifier to verify the signal presence.

Once this is done, you can start to locate the sections together. Add a layer of insulating tape across the body of the battery because the underside of the board will be making contact with it, and being metal, you don't want it to short out the connections. For the switch position, you'll want to slide the battery and board assembly inside the container, and determine the ideal position to place the switch. The switch should clear all components. Solder the wires to the switch and onto the board assembly before proceeding with the next section.

Drill a hole in the selected spot when you're sure of the final position. Because this is a PCB-mounted switch, there is none of the usual securing hardware, so it'll need to be epoxied into place. Use the two-part quick-setting type. Only a very small amount of epoxy is needed. Make sure that the epoxy does not interfere with the slide portion of the switch. The flying ground lead is attended to by drilling a hole through the container base and threading the ground cable through. Tie a knot on the inside of the wire to prevent any stress from ripping the ground solder connection away.

Finally, slide in the battery and assembly board. Make sure the probe tip section inside the lid also clears the parts. Attach the lid to complete the project. The on switch position should be marked because there's no conventional LED to indicate when the power is on.

Test Usage

For test purposes, we'll assume that you have a recently built power amplifier connected to a speaker, which is required to be tested. Switch on the power amplifier and connect the signal injector's ground lead to the power amplifier's ground connection. Turn on the signal injector and touch the probe tip to the power amplifier's input terminal. All being well, you should hear a tone coming from the speaker. This will be a harsh-sounding tone if all is well because the LM 555 is generating a square wave. This is quite normal because of the high harmonic content in a square wave. If you can't hear any signal, then you need to go back and check the power amplifier's circuitry.

The benefit of a having a dedicated signal injector is that you can quickly probe onto various points in the circuit under test and verify correct operation. It works best, of course, if you have a multistage amplifier to test because you can work backward, checking each earlier stage in turn. Circuit-wise, there is a resemblance between this signal injector and a signal generator, but the difference and the usefulness come about because of how the project is actually built and used.

Parts List

Semiconductors
IC1: LM 555 timer

Resistors
All resistors are 5 percent 1/4 W.

R1: 1 kohm
R2: 10 kohm
R3: 470 kohm
R4: 1 kohm

Capacitors
All nonpolarized capacitors are disc ceramic. All electrolytic capacitors have a 25 V rating.

C1: 0.1 µF
C2: 0.01 µF
C3: 0.1 µF

Additional Materials

S1: single pole, single throw miniature switch, printed circuit board–mounting type
Power supply: 9-volt battery

Project #12: Signal Monitor

Introduction

This is the complement of the first circuit. The signal monitor does the work of the backend monitoring when you have a test setup going. In instances where a preamplifier is under test, the signal monitor acts as the convenient block because it has its own internal speaker. All you're going to need from the signal monitor output is again some audio indication whether or not the circuit under test is functioning.

With that in mind, note that this project uses a very small speaker, 1 inch in diameter or less if you can find one, because we want to keep the physical size of the finished unit as compact as possible. Because this is a basic monitor circuit, we're leaving out any extraneous components, which would be normally included, because the signal monitor will be operating only intermittently; plus there is no special high-fidelity audio requirement from the unit (it is purely intended for use as a verifier device).

Circuit Description

Figure 18-2 shows the electrical schematic for the signal monitor. The trusty LM 386 (IC1) audio power amplifier is going to be used as our driving IC here. It's a circuit that's been cut down to include just the basic component count necessary to make it function. Only three components are used in this design: one IC and two capacitors. The incoming signal (from the device under test) is fed into the signal input (pin #2) of IC1, entering via jack socket J1, and coupling into the audio input coupling capacitor C1 (0.1 µF). The signal ground (pin #3) is returned to the circuit's common ground rail.

The audio signal output is taken from pin #5 via a coupling capacitor, C2 (100 µF), to drive a very small internal speaker. Power for the circuit is supplied from a 9-volt battery, which is fed to the Vcc terminal (pin #6). The power supply return ground goes to pin #4. A small power switch, S1, completes the component. The power amplifier has a gain of ×20, which is more than sufficient for our purpose. As in the previous project, the customary power supply components have been shorn off in this minimalist design.

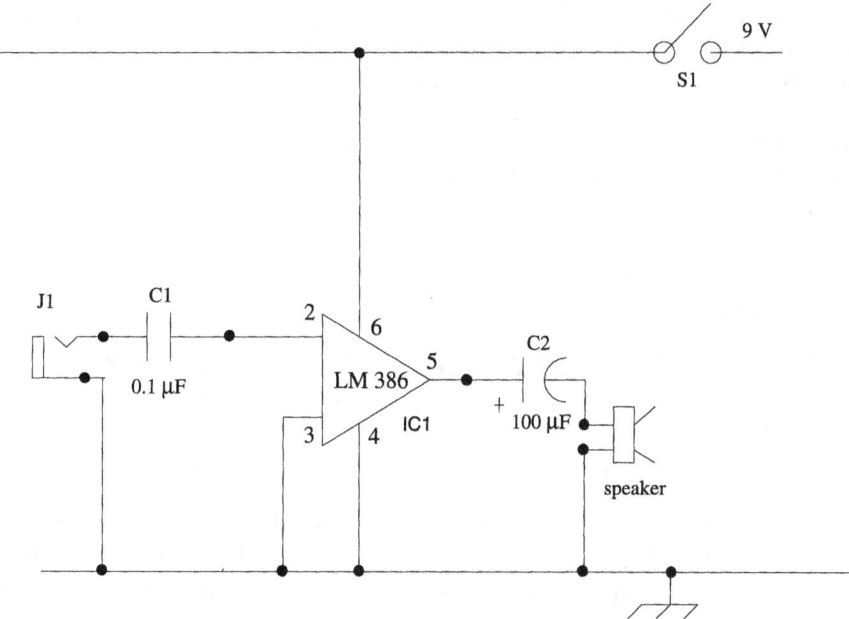

Figure 18-2 Signal monitor

Construction Details

Start with locating a suitable project case that is large enough to house the various circuit blocks. There are three key component blocks: the battery, the speaker, and the assembly board. The largest block will most likely be the speaker (depending on what you can locate). A 1-inch diameter size is a good one to aim for because it's small enough to fit a small project case. Once that has been identified, you can choose the project case to suit. It can be anything suitable and is not necessarily limited to the usual electronics project case. There won't be any undue usage, and it's not going to be seen in any case. This is a workhorse of a project.

For the input to the signal monitor, we're going to need to have two flexible flying leads coming out, with small alligator clips attached as terminations. Make sure you use a red lead (connected to input capacitor C1) for the signal input and a black lead for the ground. It's advisable to anchor these leads to the circuit board because there will be lots of stress on the leads when in use, and you don't want the solder connections ripping from the board. Small plastic ties are ideal to secure the leads to the board. Use an IC socket so you can (if needed) replace the IC. You'll need only a very small section of assembly board because so few components are needed.

The rest of the circuit connections are the power and ground lines. The board will possibly come into contact with various metal surfaces, such as the battery and speaker, so place an insulating sleeve (this could be a simple cardboard enclosure) around the completed assembly board. The circuit can be checked in situ because it's likely that your case will have a bottom and top (lid) section, thus making access easy. Mount the power-on switch, S1, wherever there is an appropriate space in the arrangement. Take care also to locate the switch terminals away from any metal areas. Choose the smallest switch you can find and mount this onto the lid section of the container of your choice.

Test Usage

Checking out this signal monitor couldn't be simpler. You've already built the previous signal injector, and we'll be coupling the two circuits together here. Just switch on the signal monitor and connect its black ground lead to the signal injector's ground lead. Clip the red (input) lead from the signal monitor to the probe (output) tip of the signal injector.

When you switch on the signal injector, you'll get a strong tone coming from the speaker, if all is well with the signal monitor. The signal injector has a 20-mV signal level. When fed into the signal monitor, which has a gain of ×20, the combined signal output is 400 mV. If no tone emerges from the signal monitor, then go through the usual component check for placement accuracy of components. Because we're using a square wave source (the LM 555 timer), the audio output is going to sound harsh as a result of the presence of signal harmonics in the square wave, but this is how it's supposed to sound.

Assume once more that you have a multistage preamplifier under test that needs to be checked. The signal monitor is going to be coupled to the last amplifier in the preamplifier chain. Make the appropriate connections with the signal monitor's red lead going to the preamplifier output terminal and the signal monitor's black lead going to the preamplifier ground. Switch on the signal monitor first and then the preamplifier under test.

When you have several series circuits running off different supplies, the rule of thumb is to always switch on the circuits in sequence, working backward, from the last to the first. With just the signal monitor and preamplifier under test switched on, no output will come from the speaker at this stage because no signal is being applied to the input.

The signal source is going to be the signal injector that was described in the previous project. Connect the signal injector's ground lead to the circuit under test, switch on the power, and, working backward starting from the last preamplifier stage, inject the signal injector's signal into the preamplifier's input terminal.

All being well, you will get a signal from the signal monitor. Depending on the gain of the amplifier under test, the output signal could well be distorted, if the final signal output is higher than the signal monitor can handle. But this result is not important because all we want to verify is that a signal is passing through the amplifier under test. In practice, all you need to do is just touch the probe tip from the signal injector to the input of the amplifier under test. That way, you won't get an ear-shattering blast from the speaker.

If no signal emerges, then that stage is suspect, so go back and troubleshoot that stage. But given that there is a signal present, then move back to the next-to-last stage, and once more couple the signal injector into the preamplifier's input stage. With each stage that you move back, the speaker signal is going to get louder. This sequence of testing by working backward essentially isolates the faulty stage (if there is one). As you can see, combining the two test units (the injector and monitor) together provides you with a versatile verifier.

Parts List

Semiconductors
IC1: LM 386 low-power audio amplifier

Resistors
All resistors are 5 percent 1/4 W.
R1: 4.7 kohm

Capacitors
All nonpolarized capacitors are disc ceramic. All electrolytic capacitors have a 25 V rating.

C1: 0.1 μF
C2: 100 μF
C3: 0.1 μF
C4: 100 μF

Additional Materials

D1: LED
Subminiature speaker, preferably 1-inch diameter size
S1: single pole, single throw miniature switch
Power supply: 9-volt battery

Index

A
ac voltage
  capacitors and, 22
  electronic circuit power source, 29
Active components, 23
Amplifiers. *See also* Preamplifiers
  alternating current type, 60
  analog electronics and, 2
  assembly techniques for, 33
  audio amplifier types, 29
  fixed low-gain audio power amplifier, 59–65
  guitar headphone amplifier, 83–89
  integrated circuits and, 23
  operational, 3
  power amplifier, 2, 29
  schematics reading for, 31, 32
  testing with signal injector, 113
  as typical electronic circuit, 29, 30
  variable-gain audio power amplifier, 71–77
Analog circuits, 23
Analog electronics overview, 1–3
  operational amplifier, 3
  temperature and, 1–2
  video versus analog, 1
Analog meters, 14
Assembly boards. *See also* SINGMIN PCB circuit assembly board
  dual-gain electret microphone audio preamplifier, 105
  fixed-frequency audio tone generator, 68, 69
  fixed-gain audio preamplifier, 77–82
  fixed low-frequency LED flasher, 48
  fixed low-gain audio power amplifier, 63
  guitar headphone amplifier, 86–87
  signal injector, 111
  signal monitor, 115–116
  variable-gain audio power amplifier, 74
  variable-gain hi/lo response audio preamplifier, 98
  variable low-frequency LED flasher/driver, 55, 56
  visual electronic metronome, 92–93
Assembly techniques, 33–35. *See also* Construction details for ten simple projects
Audio equipment
  dual-gain electret microphone audio preamplifier, 101–108
  fixed-frequency audio tone generator, 65–71
  fixed-gain audio preamplifier, 77–82
  fixed low-gain audio power amplifier, 59–65
  guitar headphone amplifier, 83–89
  variable-gain audio power amplifier, 71–77
  variable-gain hi/lo response audio amplifier, 94–101
Audio taper and potentiometers, 12
Axial lead devices, 37, 38

B

Batteries
- electronic theory and, basic, 29
- fixed low-frequency LED flasher and, 47, 48
- fixed low-gain audio power amplifier and, 60, 62
- guitar headphone amplifier and, 86
- schematic reading and, 31
- variable low-frequency LED flasher/driver and, 55

Bending resistors/capacitors, 9, 37
Bridged solder, 27–28, 34

C

Capacitors, 21–22
- determining type to use, 38
- disc ceramic type, 32, 37, 46–47
- dual-gain electret microphone audio preamplifier, 102, 104
- electrolytic capacitor, 22
- fixed low-frequency LED flasher, 44, 46
- fixed low-gain audio power amplifier, 60, 62
- fixed-frequency audio tone generator, 66, 68
- fixed-gain audio preamplifier, 77, 79
- guitar headphone amplifier, 83, 86
- measurement units, 21
- nonpolarized, 22
- property of, special, 21
- schematic reading and, 32
- signal injector, 111, 113–114
- signal monitor, 114, 117
- use for, common, 21
- variable low-frequency LED flasher/driver, 52, 54
- variable-gain audio power amplifier, 71, 73
- variable-gain hi/lo response audio preamplifier, 95, 97
- visual electronic metronome, 90, 92
- voltage rating for, 22

Car alarms, 43
Carbon microphones, 101
Circuit boards. *See also* SINGMIN PCB circuit assembly board
- assembly techniques, 33–35
- components handling, 37–38

Circuits
- digital, 4–5
- operational amplifier, 3
- schematic reading for, 31–32
- signal injector as tester, 109–114
- signal monitor, 114–117
- video, 1

Cold solder joints, 27, 34
Color coding resistors, 7–8
Components handling, 37–38
Construction details for ten simple projects, 43–108
- dual-gain electret microphone audio preamplifier, 101–108
- fixed-frequency audio tone generator, 65–71
- fixed-gain audio preamplifier, 77–82
- fixed low-frequency LED flasher, 43–52
- fixed low-gain audio power amplifier, 59–65
- guitar headphone amplifier, 83–89
- variable-gain audio power amplifier, 71–77
- variable-gain hi/lo response audio preamplifier, 94–101
- variable low-frequency LED flasher/driver, 52–59
- visual electronic metronome, 89–94

Current and Ohm's Law, 15–16, 18
Cycle definition for a sine wave, 2

D

DC voltage
- capacitors and, 22
- electronic circuit power source, 29
- polarities of, 29

Digital circuits, 23
Digital electronics overview, 3–5
- logic circuits as digital circuits, 4
- types of digital circuits, 5

Digital meters, 13, 14
Digital square wave, 4
Disc ceramic capacitors, 32, 37, 46–47
Drill, small battery-powered, 25
Dual-gain electret microphone audio amplifier, 101–108

Index

assembly board, 105
capacitors, 102, 104
circuit description, 101
component identification, 104
conclusion, 108
construction details, 105–108
integrated circuits, 104
mechanical components, 104–105
parts list, 101–102
pin connections, 102–104
resistors, 102, 104
schematics, 103, 105
semiconductor, 101

E

Electric guitars, 83–89
Electrolytic capacitors, 22
    fixed low-frequency LED flasher using, 46, 47
    schematic reading and, 32
Electronic projects. *See* Construction details for ten simple projects
Electronics theory, basic, 29–30

F

Files, needle-nosed, 25
Fixed-frequency audio tone generator, 65–71
    assembly board, 68
    capacitors, 66, 68
    circuit description, 65
    component identification, 67–68
    conclusion, 71
    construction details, 68–70
    integrated circuits, 68
    mechanical components, 68
    parts list, 66
    pin connections, 66–67
    resistors, 66, 68
    schematics, 67, 69
    semiconductor, 66
Fixed-gain audio preamplifier, 77–82
    assembly board, 80
    capacitors, 77, 79
    circuit description, 77
    component identification, 79–80
    conclusion, 82
    construction details, 81–82
    integrated circuits, 79
    mechanical components, 80
    parts list, 77–78
    pin connections, 78–79
    resistors, 77, 79
    schematic, 78
    semiconductor, 77, 80
Fixed low-frequency LED flasher, 43–52
    assembly board, 48
    capacitors, 46–47
    circuit description, 43
    component identification, 45–47
    conclusion, 51–52
    construction details, 49–51
    integrated circuits, 47
    mechanical components, 47–48
    parts list, 43–44
    pin connections, 44–45
    resistors, 46
    schematic, 45
Fixed low-gain audio power amplifier, 59–65
    assembly board, 63
    capacitors, 62
    circuit description, 60
    component identification, 62
    conclusion, 65
    construction details, 63–65
    integrated circuits, 62
    mechanical components, 62
    parts list, 60
    pin connections, 60–62
    resistors, 62
    schematics, 61, 63
Frequency
    definition, 2
    fixed-frequency audio tone generator, 65–71
    fixed low-frequency LED flasher, 43–47
    variable low-frequency LED flasher/driver, 52–59

G

Gain
    dual-gain electret microphone audio preamplifier, 101–108
    fixed-gain audio preamplifier, 77–82
    fixed low-gain audio power amplifier, 59–65

Gain—*continued*
  variable-gain audio power amplifier, 71–77
  variable-gain hi/lo response audio preamplifier, 94–101
Glue gun, 25
Ground connection, common, 30
Ground rail and circuit schematics, 31
Guitar headphone amplifier, 83–89
  assembly board, 86–87
  capacitors, 83, 86
  circuit description, 83
  component identification, 85–86
  conclusion, 89
  construction details, 87–89
  integrated circuits, 86
  mechanical components, 86
  parts list, 83–84
  pin connections, 84–85
  resistors, 83, 85
  schematic, 86
  semiconductor, 83

H
Headphone amplifier, guitar, 83–89. *See also* Guitar headphone amplifier

I
Injury prevention, 25, 26
Integrated circuits (ICs), 23–24
  categories of, 23
  components, 23
  dual-gain electret microphone audio amplifier and, 104
  fixed-frequency audio tone generator and, 68
  fixed-gain audio preamplifier and, 79
  fixed low-frequency LED flasher and, 47
  fixed low-gain audio power amplifier and, 62
  functions of, common, 23
  guitar headphone amplifier and, 83, 86
  handling, 38
  IC sockets, 41–42
  schematic reading and, 32
  signal monitor, 114
  types of, three, 23
  variable-gain audio power amplifier and, 73
  variable-gain hi/lo response audio preamplifier and, 98
  variable low-frequency LED flasher/driver and, 54
  visual electronic metronome and, 92

K
Kilo definition, 7

L
Light-emitting diodes (LEDs), 17–18
  advantages of, 17
  fixed-frequency audio tone generator and, 66
  fixed low-frequency LED flasher, 43–52
  fixed low-gain audio power amplifier and, 60, 62
  schematics reading and, 31–32
  use for, 17
  variable low-frequency LED flasher/driver, 52–59
  visual electronic metronome and, 89, 90
Linear taper and potentiometers, 12
Logic circuits, 4–5

M
Metronome, visual electronic, 89–94. *See also* Visual electronic metronome
Microfarad (μF), 21
Microphones. *See* Dual-gain electret microphone audio preamplifier
Multimeters, 13–14
  definition, 13
  function, 13
  LED current flow measurement by, 18
  resistance range and, 46, 55
  SPDT switch and, 20
  switches and, 19
  types of, 13

N
Needle-nosed pliers, 25, 37
Negative terminal, 29, 30, 31
Nonpolarized capacitors, 22

O

Ohm's Law, 2, 15–16, 18
Operational amplifiers (op-amps), 3
Oscillators
  integrated circuits and, 23
  signal output of, 29
  square wave, 65
Oscilloscope
  definition, 3
  digital signals and, 4

P

Picofarad (pF), 21
Pliers, 25, 37
Poles and switches, 19
Positive terminal, 29, 30, 31
Potentiometers, 11–12
  frequency variation provided by, 52, 55
  linear and audio taper and, 12
  potential dividers as another name for, 11
  terminals of, 11
  values of, typical, 11
  as volume control function, 74
Power amplifier, 2, 29
Preamplifiers, 2
  dual-gain electret microphone audio preamplifier, 101–108
  fixed-gain audio preamplifier, 77–82
  function of, 29
  power amplifier versus, 2
  testing with signal monitor, 116
  variable-gain hi/lo response audio preamplifier, 94–101

R

Radical lead devices, 37, 38
Reamer, 25
Resistors, 3, 7–10. *See also* Multimeters; Potentiometers
  bending, 9, 37
  connecting, 9–10
  dual-gain electret microphone audio preamplifier, 101, 102, 104
  fixed-frequency audio tone generator, 66, 68
  fixed-gain audio preamplifier, 77, 79
  fixed low-frequency LED flasher, 44, 46
  fixed low-gain audio power amplifier, 60, 62
  guitar headphone amplifier, 83, 85
  measuring, 7
  power rating of, 8
  range availability, 9
  resistance value color code, 7–8
  signal injector and, 111, 113
  signal monitor and, 117
  use for, common, 8
  variable-gain audio power amplifier, 71, 73
  variable-gain hi/lo response audio preamplifier, 95, 97
  variable low-frequency LED flasher/driver, 52, 54
  visual electronic metronome, 90, 92
Rosin core solder, 25

S

Safety guidelines, 25, 26
Schematics
  for dual-gain electret microphone audio preamplifier, 103, 105
  for fixed-frequency audio tone generator, 67, 69
  for fixed-gain audio preamplifier, 78
  for fixed low-frequency LED flasher, 45
  for fixed low-gain audio power amplifier, 61, 63
  for guitar headphone amplifier, 86
  reading, 31–32
  for variable-gain audio power amplifier, 72, 74
  for variable-gain hi/lo response audio preamplifier, 96, 98
  for variable low-frequency LED flasher/driver, 53, 56
  for visual electronic metronome, 92
Screwdrivers, 25
Semiconductors
  complementary metal oxide (CMOS), 5
  dual-gain electret microphone audio preamplifier, 101

Semiconductors—*continued*
   fixed-frequency audio tone generator, 66
   fixed-gain audio preamplifier, 77, 80
   fixed low-frequency LED flasher, 43
   fixed low-gain audio power amplifier, 60
   guitar headphone amplifier, 83
   signal injector, 113
   signal monitor, 117
   variable-gain audio power amplifier, 71
   variable-gain hi/lo response audio preamplifier, 95
   variable low-frequency LED flasher/driver, 52
   visual electronic metronome, 90, 91
Signal injector, 109–114
   assembly board, 111
   capacitors, 111, 113–114
   circuit description, 110–111
   construction details, 111–113
   introduction to, 109–110
   parts list, 113–114
   resistors, 111, 113
   schematic, 110
   semiconductor, 113
   test usage, 113
Signal monitor, 109, 114–117
   assembly board, 115–116
   capacitors, 114, 117
   circuit description, 114–115
   construction details, 115–116
   integrated circuits, 114
   introduction to, 114
   parts list, 117
   resistors, 117
   schematic, 115
   semiconductor, 117
   test usage, 116–117
Sine waves, 25
SINGIN PCB circuit assembly board, 39–42
   diagrams of, 40
   voltage supply rails diagram, 41
Single pole double throw switch (SPDT), 19–20, 53, 102, 104
Single pole single throw switch (SPST), 19, 44, 47, 53, 60, 72, 78, 84, 90, 95, 102
Soldering, 27–28, 34–35
   errors in, 34, 35
   good practice for, 34
   suggestions for, 41–42
   wires used for, 33
Soldering irons, 25, 26, 27, 34
Solid hook-up wire, 33
Stranded hook-up wire, 33
Supply voltage definition, 30
Switches, 19–20
   purpose of, 19
   single pole double throw switch type of, 19–20, 53, 102, 104
   single pole single throw switch type of, 19, 44, 47, 53, 60, 72, 78, 84, 90, 95, 102

T

Throws and switches, 19
Tools, types and uses, 25–26
Troubleshooting test equipment, 109–117
   signal injector, 109–114
   signal monitor, 109, 114–117

V

Variable-gain audio power amplifier, 71–77
   assembly board, 74
   capacitors, 71, 73
   circuit description, 71
   component identification, 73
   conclusion, 77
   construction details, 74–76
   integrated circuits, 73
   mechanical components, 74
   parts list, 71–72
   pin connections, 72–73
   resistors, 71, 73
   schematics, 72, 74
   semiconductor, 71
Variable-gain hi/lo response audio preamplifier, 94–101
   assembly board, 98
   capacitors, 95, 97
   circuit description, 95
   component identification, 97–98

conclusion, 101
construction details, 99–101
integrated circuits, 98
mechanical components, 98
parts list, 95
pin connections, 96–97
resistors, 95, 97
schematics, 96, 98
semiconductor, 95
Variable low-frequency LED
    flasher/driver, 52–59
  assembly board, 55
  circuit description, 52
  component identification, 54
  conclusion, 59
  construction details, 55–59
  integrated circuits, 54
  mechanical components, 55
  parts list, 52–53
  pin connections, 53–54
  schematics, 53, 56
Video circuits, 1
Visual electronic metronome, 89–94
  assembly board, 92–93
  capacitors, 90, 92
  circuit description, 90
  component identification, 91–92
  conclusion, 94
  construction details, 93–94
  integrated circuits, 92
  mechanical components, 92
  parts list, 90
  pin connections, 90–91
  resistors, 90, 92
  schematic, 92
  semiconductor, 90, 91
Voltage
  capacitors and, 21–22
  electronic theory and, basic, 29
  LEDs and, 17
  Ohm's Law and, 15–16
  overloads and multimeters, 14
Voltmeter, 47

W
Wire
  nicked, 33
  solid hook-up wire, 33
  stranded hoop-up wire, 33
  stripping insulation from, 33
Wire cutters, 25
Wire stripper, 25, 33
Wrenches, 25

Z
Zobel network, 71

www.ingramcontent.com/pod-product-compliance
Lightning Source LLC
Chambersburg PA
CBHW051444290426
44109CB00016B/1674